Harnessing Hydrogen

The Key to Sustainable Transportation

James S. Cannon
Editor: Sharene L. Azimi

INFORM, Inc.
120 Wall Street
New York, NY 10005-4001
Tel 212 361-2400
Fax 212 361-2412
E-mail Inform@igc.apc.org

Printed in the United States of America

Library of Congress Cataloging-in-Publications Data

Cannon, James Spencer, 1949-
Harnessing hydrogen : the key to sustainable transportation /
James S. Cannon.
p. cm.
Includes bibliographical references and index.
ISBN 0-918780-65-9 : $30.00
1. Automobiles – Motors (Compressed-gas) 2. Hydrogen as fuel.
3. Fuel cells. I. Azimi, Sharene L. II. Title.
TL229.H9C36 1995
629.25'38 – dc20 95-341
CIP

Printed on recycled paper

INFORM, Inc., is a nonprofit environmental research organization that examines business, institutional, and government practices that threaten our environment and public health; assesses changes business and government are making to improve their performance; and identifies new business strategies and technologies moving the United States toward an environmentally sustainable economy. INFORM's research currently focuses on strategies to reduce industrial and municipal waste and to preserve air and water quality.

A copy of INFORM's Annual Report may be obtained by contacting the Office of Charities Registration, 162 Washington Avenue, Albany, NY 12231, or INFORM, Inc., 120 Wall Street, New York, NY 10005-4001.

Contents

Acknowledgments

The author wishes to thank literally hundreds of people within the international hydrogen energy community who made valuable contributions to this report, particularly those who agreed to be interviewed or who published professional papers, studies, and books concerning hydrogen vehicles during the research phase of the project. Special thanks go to the editors of three publications that report to the world key developments in the rapidly changing and expanding arena of alternative transportation fuels – Jerry Sinor, publisher of *Clean Fuels Report;* Peter Hoffman, editor of *Hydrogen & Fuel Cell Letter;* and T. Nejat Veziroğlu, editor-in-chief of *International Journal of Hydrogen Energy.*

Once the manuscript was written, several reviewers painstakingly dissected the text, uncovering errors and freely offering their own insights and suggestions. The reviewers included Roland Hwang and Jason Mark at the Union of Concerned Scientists, Robert Mauro at the National Hydrogen Association. Sandy Thomas of Directed Technologies, Inc., Jay Keller from Sandia National Laboratories, Catherine Gregoire Padró at the National Renewable Energy Laboratory, and Robert Socolow and Margaret Steinbugler of the Center for Energy and Environmental Studies at Princeton University also reviewed portions of the final draft.

I would like to thank many people at INFORM whose assistance, guidance, and advice made possible the completion of *Harnessing Hydrogen: The Key to Sustainable Transportation.* INFORM's president, Joanna Underwood, as always was the mainstay of support and encouragement throughout the project. Sharene Azimi, the project editor, ably worked for more than one year on multiple drafts of the report, skillfully molding ideas and transforming even the most incomprehensible prose into what I hope is a coherent manuscript. Elisa Last, production coordinator, is responsible for the creative design of the book and, along with Mimi Bluestone, INFORM's publications director, its final publication.

INFORM thanks the many individuals and organizations that provided general support; Dr. Dorcas McDonald for her support of this project; and the following foundations whose grants made this study possible: The German Marshall Fund of the United States, The Joyce Mertz-Gilmore Foundation, The Pew Charitable Trusts, the Helen Sperry Lea Foundation, the Surdna Foundation, Inc., and the Mark and Catherine Winkler Foundation.

Finally, on a personal note, I wish to express my gratitude to my wife, Cheryl Stevenson, and to my son, David, for their patience, interest, and good humor, and for countless everyday actions that bring joy into my life.

While this report benefited from the help of these individuals and organizations, the content and analysis are the sole responsibility of INFORM.

Preface

For the United Nations Development Programme (UNDP), sustainable human development means focusing resources on four key areas: eradicating poverty, increasing women's role in development, providing people with income-earning opportunities, and protecting and regenerating the environment.

Initiatives in the transportation and energy sectors are important to achieving sustainable human development. Transportation and energy-service needs evolve and expand as countries develop. Because of these sectors' tremendous economic and environmental impact, any development strategy should take them into account. If current patterns of transportation energy production, distribution, and consumption continue, progress in a number of countries could slow dramatically or even come to a halt. Industrialized and developing countries must reconsider their use of transportation and energy. Minor adjustments to conventional transportation and energy systems will not meet the challenge. A major shift away from business-as-usual is needed.

"Agenda 21," an action agenda issued at the conclusion of the United Nations Conference on Environment and Development, called for more efficient systems for producing, distributing, and consuming energy, and for greater reliance on environmentally sound transportation and energy systems, with special emphasis on renewable sources of energy. UNDP is helping countries implement national energy policies that support their development strategies. This initiative is demonstrating how the energy sector can be a tool for development by giving people income-earning opportunities, building up government institutions' capacities for protecting the environment and increasing transportation and energy efficiency, and accelerating technological development. UNDP continues to formulate new ways to address these important issues.

Harnessing Hydrogen looks at the long-term goal of sustainable transportation and at the potential of hydrogen to serve as the backbone of a sustainable transportation system of the 21st century. Today, gasoline and diesel fuel 99 percent of the world's 520 million vehicles. The burning of fossil fuels is a major contributor to some of our most troubling public health and environmental problems, including respiratory disease, smog, acidification, and global warming. The fuel and automotive industries have made significant progress in reducing certain tailpipe emissions. But complex automotive filtration systems still do nothing to prevent carbon dioxide emissions — a by-product of fossil fuel burning thought to be contributing to a potentially disastrous warming of the atmosphere.

With the 1990 Clean Air Act and the 1992 Energy Policy Act, the United States recognized the critical importance of starting to make a long-term transition to cleaner transportation fuels. Today it is clear that we must do more: ultimately, we must find transportation fuels compatible with sustainable development. The challenge is to find alternatives that can be used virtually pollution-free.

This publication argues that hydrogen, as a transportation fuel, is demonstrated technology that can serve us tomorrow if nurtured today. The choices we make and the actions we take now will have enormous environmental and economic implications for our future. A wide variety of technologies are available or emerging for the production of hydrogen. One of the most significant actions we can take today is to expand our use of natural gas vehicles because natural gas can serve as a bridge to the development of hydrogen technologies and infrastructures.

Along with better mass transit systems, hydrogen technologies will be needed to meet the needs of expanding markets in the industrialized and developing societies. We must find ways to help developing nations, which are not already fully wedded to the internal combustion engine and a gasoline distribution system, develop such sustainable transportation and energy systems.

James Gustave Speth
Administrator
United Nations Development Programme

Chapter 1: Introduction

From Horses to Hydrogen

The world is on the cusp of a revolution in the production and use of energy for transportation. In this revolution, a world almost totally dependent on polluting and rapidly depleting oil supplies will be left behind. Current trends in fuel use indicate that this oil-dependent system will be supplanted by a transportation system that favors cleaner fuels such as natural gas for the present and that aims to use pollution-free fuels derived from renewable resources over the longer term. As patterns in transportation fuel use change, technological innovations in highly efficient and nonpolluting electrically driven motors will offer new commercial engines and propulsion systems capable of replacing the wasteful internal combustion engine that has dominated the 20th century automotive industry. Together, new fuels and the engines to use them will combine to produce a new transportation era built entirely on sustainable energy.

Because hydrogen can be produced cleanly from renewable energy resources, and because its own use is virtually without pollution, it may provide the long-term solution to the problems created by dependence on oil as the almost exclusive source of automotive fuels worldwide. A plausible strategy for the transition to hydrogen could involve the initial replacement of conventional automotive fuels, gasoline and diesel, with natural gas, followed by the gradual introduction of hydrogen, a gaseous fuel in many ways similar to natural gas. This pathway moves from a limited and polluting fossil fuel, through a cleaner and more abundant fossil fuel, ultimately arriving at a sustainable and pollution-free energy system. Although it is not a foregone conclusion that these transitions will occur, the natural gas-to-hydrogen bridge has many advantages and deserves much more serious examination than it has received in energy and environmental policy deliberations to date.

It may take decades for the transition to hydrogen energy now underway to take hold and the better part of a century to complete them. But much of the new ground will be broken in the next decade as future players make their mark on transportation for the 21st century, much as the Henry Fords and the John D. Rockefellers laid claim 100 years ago to the technologies and resources that largely shaped transportation in this century.

Global Energy Transitions

The evolution to a sustainable energy resource may begin first and proceed most quickly in transportation, where the world's energy and environmental problems are particularly extreme. A drive toward a sustainable energy economy, however, applies to other uses of energy as well. In fact, the transportation revolution from horses a century ago to hydrogen-powered vehicles in the 21st century is occurring within the context of broader energy transitions in the patterns of global energy use. The movement toward sustainable energy in transportation that is now under way is part of the second major energy transition in human history. The first transition began two centuries ago, when the commercialization of coal marked the gradual abandonment of thousands of years of exclusive reliance on renewable resources, mostly food and wood. This first energy transition did not affect individual transportation until even more recently, when the discovery of oil as a fuel in the mid-1800s spurred the gradual replacement of 20 million horses in the United States with automobiles.[1] Now, as the world reaches a peak in fossil fuel use, the cycle is turning to renewable sources of energy that humans were previously unable to harness.

We Are Not Just What We Eat

In 1492, when Columbus first sailed across the Atlantic Ocean, virtually all the energy used by the world's 400 million people came from renewable sources. Technologically, it was a simpler world then, and most energy was released in the form of human exertion and the labor of draft animals, mainly horses and oxen. Food, a renewable resource, provided this energy and was the principal energy resource in Columbus' time. The burning of wood, a renewable resource as long as the growth of new trees equals or exceeds the amount cut, provided most of the other energy used in 1492. Additional renewable resources included hydropower harnessed by creek-side water wheels and wind energy captured and put to work, for example, by the sails of Columbus' ships. Nonrenewable fossil fuels – coal, oil, and natural gas – were as yet largely undiscovered, dormant in the underground vaults where they evolved millions of years ago.[2]

The world is a startlingly different place today. A tenfold increase in population since the time of Columbus, to 5.6 billion people in 1994, has been accompanied by a more than hundredfold jump in energy use.[3] World energy use now stands at about 385 quadrillion British thermal units (Btu).[4] One Btu is the amount of energy required to raise the temperature of one pound of water by one degree Fahrenheit, or about the energy released from burning one kitchen match. A quadrillion is equal to a billion million. An annual energy use of 385 "quads" of Btu, therefore, is about equal to burning 385 billion piles of matches, with one million matches stacked in each pile.

It is, of course, not matches that the world burns to provide the energy it needs. Today, it is primarily fossil fuels. Insignificant energy providers in 1492, they now account for the lion's share, about 80 percent, of the world's energy supply.[5] Most fossil fuel use has occurred in this century, marking one of the most significant developments in human history. Among other things, it has meant that food has been superseded by other energy resources not only capable of doing work for people, but able to do much more of it. A pound of coal or a "glass" of gasoline contains more energy than all the food eaten daily by an average human. The energy content of the food digested by the world's population today is less than 20 quads of

Btu per year.[6] Even including the work of draft animals, the energy released by food digestion is now of only minor significance compared with the total energy consumed in all human activities.

The ability to do work is no longer largely restricted to the individual strength of human beings and animals. Energy released by fossil fuels does most of the work for us. It is this uncoupling of world energy supply from food that has created the world as we know it today. As this uncoupling has taken on greater dimensions, so has the distinction between modern life and human life in previous centuries. Energy has proven to be an essential tool, transforming the societies of the industrialized countries in the 20th century. Per-capita energy consumption in the industrialized world, with its cars, air conditioners, and electronic devices, is 11 times greater than the energy used by an average person living in a developing country.[7] The United States alone, with less than 5 percent of the world's population, consumes 24 percent of its energy.

Transportation provides a compelling example of how increased fossil fuel use has enabled a lifestyle inconceivable with the energy supplies of previous eras. The average American compact car is propelled by a 100-horsepower engine. One horsepower, although technically defined as about 2,580 Btu of energy expended per hour, has as its origin the estimated force exerted by a pulling horse. Since there are 190 million vehicles on American roads today, the total pulling power is about equal to 19 billion horses – nearly 1,000 times the power of the horses used for transportation in 1900.

At the end of the 19th century, when fossil fuel use was nascent and horses were still the dominant mode of transportation, an estimated 20 million horses in the United States transported humans and market goods.[8] One-quarter of the farmland in the United States was used to grow food to feed these horses.[9] On a typical day in New York City, the streets were awash with 60,000 gallons of horse urine and buried beneath 2.5 million pounds of manure. Tetanus, dysentery, and respiratory infections were among the serious urban health problems, especially for city children, stemming from unsanitary conditions created by horses. The annual cleanup cost of the mess, plus the removal of the 15,000 horses who yearly died on the streets, approached $100 million.[10] Indeed, automotive advocates at the turn of the century included the promise of reduced pollution among the quality-of-life improvements that were widely anticipated from automobiles.

Imposing a horse-based transportation system onto today's world would be preposterous. In the first place, there is not nearly enough land available to grow the food the horses would eat. Second, federal environmental standards limit automotive tailpipe emissions of all regulated pollutants to less than 5 grams per mile driven. On a per-mile-walked basis, a horse emits from its analogous points 640 grams of solid waste and 300 grams of liquid waste.[11] The specter of Americans trying today to move about on 19 billion horses is as ludicrous as the thought of thousands of winged creatures like Pegasus propelling an airplane across the sky.

The Fossil Fuel Era

The pervasiveness of fossil fuels in modern life masks the fact that the fossil fuel era is a very recent occurrence in human history, and one that is destined to be relatively short-lived. As late as 1950, food accounted for more than two-thirds of all energy use, with wood-burning accounting for most of the rest.[12] Oil, coal, and natural gas – the three fossil fuels – provide 36 percent, 23 percent, and 19 percent of the world's primary energy supply (see Figure 1). About 6 percent of the remainder comes from uranium, a nonfossil but similarly nonrenewable energy resource. Only 16 percent comes from renewable energy, mostly in the form of wood and other biomass (10 percent) and hydropower (6 percent). Other renewable resources, including the direct use of solar and wind energy, are not yet significant enough to earn a discernible piece of the energy supply pie.

Fossil fuels were formed mostly during a 65 million-year period that ended about 250 million years ago. Massive quantities of organic plant and animal life accumulated in low-lying areas and were gradually covered by sediments and oceans. As their depth of cover increased, the organic debris became subject to high temperatures and pressures that are characteristically found deep within the Earth. Under these conditions, the organic material gradually turned into coal, oil, and natural gas. It takes 3,000 to 9,000 years for organic material to show the first signs of trans-

Figure 1. World Energy Use, 1992: 385 quadrillion Btu

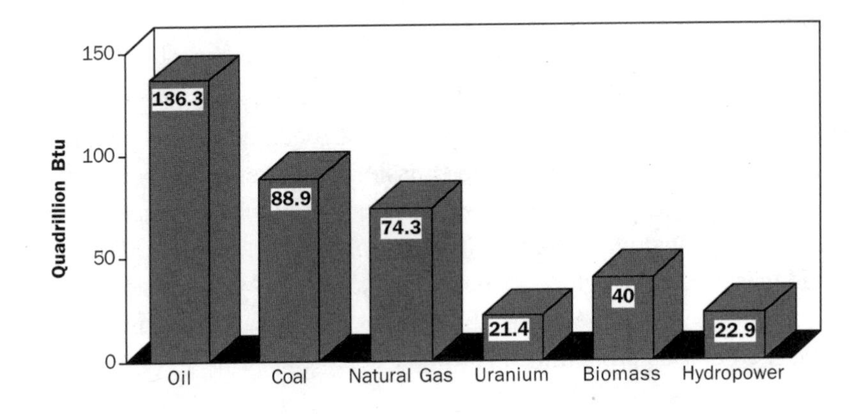

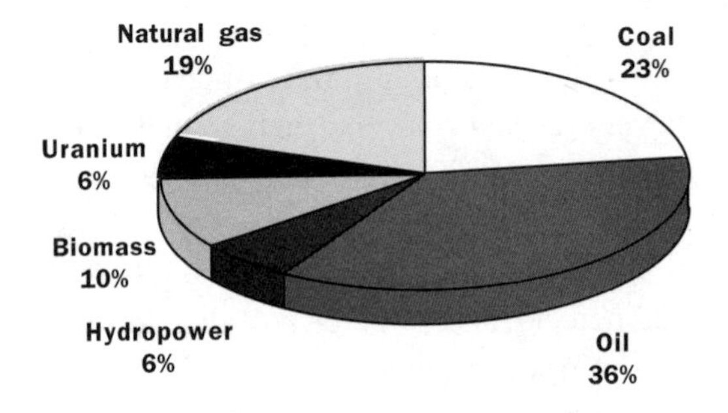

Sources: US Energy Information Administration, *International Energy Annual 1992*. Biomass data from *British Petroleum Statistical Review of World Energy*, 1989.

formation into fossil fuels, and several million years to complete the process. Fossil fuels are being burned at a rate 100,000 times greater than the rate at which they can be regenerated by natural processes. Once burned, fossil fuels are not replaced for eons, which means they are nonrenewable, depletable energy resources.[13]

The age of fossil fuel use is just a flicker in geological time and barely a wave in the tide of human history. In a striking example of this century's high level of consumption, more than 75 percent of all the fossil fuels that have been used throughout American history have been consumed in the last 100 years. Until 1900, fossil fuel consumption in the United States was less than 5 percent of today's level. It took until the 1940s for oil production to reach 50 percent of today's yield. Natural gas production did not grow to half of today's output until the mid-1950s.[14]

Looking into the future, it is likely that 75 percent of all the fossil fuels that will ever be used from now on will be burned in the next 100 years. Human dependence on fossil fuels will probably span less than 200 years – from the late 19th century to sometime in the 21st century – and the total range of any significant fossil fuel use could be less than 400 years.[15]

As we will discuss in Chapter 3, there are four major reasons for the inexorable decline in oil use: limited supplies, environmental complications, eroding economics, and destabilizing social and political repercussions. These issues apply to the other fossil fuels – coal and natural gas – to a greater or lesser degree as well. Although there are widely differing opinions about when and how the world must wean itself from reliance on fossil fuels for most of its energy, there is no dispute about the fact that the fossil-fuel era involves a one-time harvest of depletable resources that must come to an end. Furthermore, the transition from fossil fuels must occur sooner rather than later if people in the world's developing nations are to enjoy access to the quantity of energy that helps produce the quality of life Americans take for granted.

The Post-Fossil Fuel Era

Over the long term, the world will have to abandon its reliance on fossil fuels in favor of renewable energy resources. The trick is to stage this shift to renewable energy without also reimposing the quality-of-life constraints prevalent in the pre-fossil fuel era. In other words, new and greater forms of renewable energy, beyond food and wood, are needed. An imaginary model for global energy use may be useful in visualizing the world's move into and out of the fossil fuel age.[16]

The first phase of this model, the pre-fossil fuel era, stretches from the beginning of recorded history until about 1850 and is marked by low per-capita energy consumption and reliance on renewable energy resources. This energy economy was sustainable because natural processes replenished energy supplies as quickly as they were consumed. Quality of life, however, was limited by the absence of modern conveniences such as automobiles and electrical appliances.

The second phase, the fossil fuel era, has been characterized to date by the rapid increase in energy consumption and the dominance of fossil fuel energy resources, which have outstripped renewable resources in the world energy equation. World energy use has increased more than fiftyfold this century, with nearly all this growth supplied by fossil fuels.[17] As energy-consuming technologies have proliferated during this era, the quality of life of the industrialized world as we know it has emerged.

The world is currently at some point near the top of the upswing in fossil energy use. The end is in sight as the interplay of the forces discussed in Chapter 3 increases the pressure to reduce fossil fuel use. The actual date of this turnaround is debatable: For example, those most concerned about global warming believe that today's level of fossil fuel use has reached a maximum for Earth to sustain. These groups call for reductions in fossil fuel use now in order to avoid environmental catastrophe. On the other side, many energy producers believe today's fossil fuel use will vastly increase before a peak is reached sometime late next century.

A 1992 energy model developed at the Clean Energy Research Institute at the University of Miami took a middle position. According to the study, although fossil energy use would peak before the middle of the next century, another tripling of fossil fuel use, much of it provided by natural gas, could occur in the interim. Total fossil fuel use would continue to rise, but the model predicts that the contribution of fossil fuels as a percentage of the world's total energy will begin to drop early in the 21st century as increasing amounts of energy are obtained from renewable resources, mainly from solar energy. From this analysis, the modelers conclude that: "Historically, the fossil fuels era could be considered as a short interlude between a solar energy past and the solar future."[18]

Regardless of one's perspective, the limitations on fossil fuel use will, at some point, become overwhelming, and an inevitable decline in the amount of fossil fuels consumed will begin. Although fossil fuel use will slope downward in this era and eventually disappear, total energy use is likely to remain nearly constant so that quality of life can be maintained.

The gap in energy demand will be filled by renewable energy, which will gradually return as the dominant source of energy in the world.

Features of Current Global Energy Transitions

As the fossil fuel era has progressed, different fuels have dominated world energy supply at different times. The trends in fuel use during this era suggest that the world has been moving in the general direction of using cleaner fuels and has already begun to increase reliance on nonfossil fuel energy resources. The changing patterns in energy use in the 20th century are characterized by:

- Upswings and downswings in the use of specific fossil fuels over time.
- A trend toward fuels containing less carbon.
- A trend toward fuels that contain more hydrogen and are therefore more powerful and cleaner.

As society has exploited first coal, then oil, then natural gas, each successive fuel has contained more hydrogen and less carbon – making it both more powerful and cleaner than its predecessor.

Growth and Decline of Fuel Choices

Fossil fuel use has predominated over energy from renewable resources since the late 1800s. But this predominance has not been unidimensional. The overall increase in fossil fuel use has been marked by distinct periods of growth and decline in the use of specific fuels. Caesar Marchetti, working at the Austrian International Institute of Applied Systems Analysis in the late 1970s, developed a model that breaks down energy consumption by fuel source during the fossil-fuel era. His analysis of the overall trends in energy use throughout human history reveals broad patterns of change that are neither random nor wildly fluctuating.[19]

Figure 2. Energy Supply Evolution: 1850 – 2050

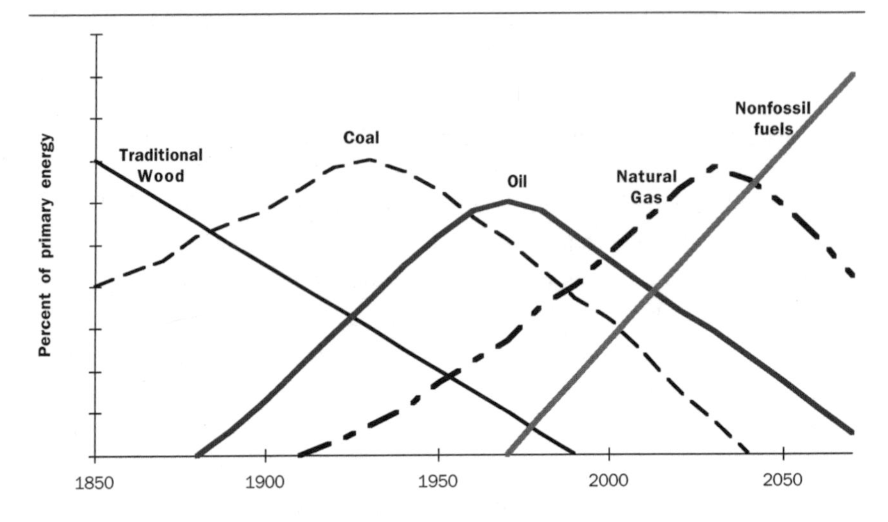

Source: Adapted from the work of Caesar Marchetti, Austrian International Institute of Applied Systems Analysis, late 1970s.

Recent world history, Marchetti has shown, has been marked by clearly delineated progressions in energy resources – from wood to coal to oil and, increasingly, to natural gas and nonfossil fuels. The patterns, based on actual past energy use and mathematical projections of the future, are shown in Figure 2. The left line maps the decline in the use of wood from the mid-1800s, when wood-burning accounted for about 70 percent of the world's primary energy consumption, through the end of this century.

The three mountain-shaped curves chart the rise and eventual decline in the use of the three fossil fuels. Coal use reached a peak in the early part of this century, when it accounted for nearly 75 percent of the world's energy. Although coal remains abundant, its decline this century has occurred largely as a result of increased availability of denser and more potent oil. Marchetti's model predicts that oil use is now reaching a peak in terms of the percentage of the world's energy it supplies. Natural gas, even cheaper and more energy-dense, is now displacing oil in the same manner that oil earlier displaced coal. According to Marchetti, natural gas use will reach a peak about 2020. The energy transitions over the past century have, by and large, not been attributable to resource depletion, but rather to increased availability of better energy resources. Since Marchetti's work, analyses of actual data have shown that world energy use closely conforms to the curves he calculated.[20]

The last line shows the appearance and growth of the next major fuel supplying the world's energy needs, one that is not based on any fossil fuel reservoir. Marchetti, writing in the 1970s, assumed much of this energy source to be energy generated at nuclear power stations; the graph plots the advent of nuclear power beginning about 1970.

While some may argue that nuclear power is a viable technology for electrical generation, INFORM has not considered an expanded role for nuclear power in this study for two reasons. First, the processes of mining transporting, using, and disposing of radioactive elements pose a serious risk of long-term damage to the environment and public health. Second, while the supply of uranium is sufficient to support a large nuclear industry, it is nonetheless a finite resource. On the other hand, hydrogen and electricity produced from renewable resources hold vast potential to move the world's energy patterns toward the nonfossil future predicted by Marchetti – without reliance on nuclear power.

Moving Away from the Carbon Atom

A key feature of the recent energy transitions has been the trend toward fuels containing less carbon in their chemical compositions. As shown in Table 1, dried wood is nearly pure carbon, containing only 10 percent hydrogen atoms. The percentage of hydrogen atoms in the fossil fuels increases from 38 percent in coal to 64 percent in oil and 80 percent in natural gas. There is no carbon in pure hydrogen.[21]

Carbon serves two functions in these fuels. First, it acts as an anchor holding these resources within human grasp. The atomic weight of carbon is 12, compared with 1 for hydrogen. The more carbon in a molecule, the greater its molecular weight. The heaviest, most carbon-rich fuels – wood and coal – are tied to the earth as solids. Lighter, less carbon-rich oil is a liquid, free to seep into the earth's caverns and surfaces, but too heavy to

Table 1. Moving Away from Carbon toward Lighter Fuels

Fuel Type	Physical State	% Carbon	% Hydrogen
Dry wood	solid	90%	10%
Coal	solid	62%	38%
Oil	liquid	36%	64%
Natural Gas	gas	20%	80%
Hydrogen	gas	0%	100%

Note: The atomic weight of carbon is 12, compared with 1 for hydrogen.

Source: *International Journal of Hydrogen Energy* articles by K. De Jong and M. Van Wechem, June 1995; and C. Marchetti, April 1985.

leap en masse into the atmosphere. Natural gas, with only one carbon atom linked to four hydrogen atoms, is so light that it naturally occurs as a gas. Lighter still, hydrogen is a gas so buoyant that most of it that is naturally occurring has escaped the atmosphere – floating into interstellar space.

The logistics of producing and distributing fuels in different physical states formed part of the reason for the succession in fuel use. Wood is easy to harvest, and digging coal out of the ground is straightforward, even with the technology that was available in the 19th century. The discovery of oil, however, awaited advances in well-drilling technology that did not occur until the mid-19th century. Oil was originally distributed in wooden barrels, which encased the liquid in a solid shell that could be transported much like wood or coal. The common unit of measurement of oil is still the barrel (equal to 42 gallons), even though oil barrels themselves have long since been relics of the past. Large-scale oil distribution did not begin in earnest until the development of pipeline technology capable of moving liquids long distances. The first oil pipeline was built in 1879 between Pennsylvania oil fields and rail terminals in Ohio.[22] Oil barges and ocean-going oil tankers followed. The United States alone has about 200,000 miles of oil pipelines.[23]

Producing and moving gaseous fuels presented even more technological challenges. In fact, natural gas produced simultaneously during oil production was routinely vented to the atmosphere through the 1930s because technology did not exist to capture the gas, compress it, and distribute it to market in a cost-effective manner. Although largely eliminated in the United States, flaring and venting of natural gas continues today, especially in remote areas. High-pressure natural gas pipeline technology did not become widespread until the middle of this century, permitting the growth in natural gas use that is now occurring. The network of natural gas pipelines in the United States now exceeds 1.3 million miles.[24] Natural gas liquefaction technology and transportation in specially insulated tankers are more recent developments, reflecting even more complex technology. Systems for storing and distributing natural gas are also capable of being modified to handle the other gaseous fuel – hydrogen.

More Hydrogen: The Trend toward More Powerful Fuels

The second role of carbon, besides serving as an anchor, is to provide energy when burned. However, it provides much less energy than hydrogen. The energy content of fuels increases as the percentage of carbon decreases in favor of hydrogen; breaking chemical bonds involving hydrogen releases more energy than breaking chemical bonds involving carbon. While the term "hydrogen energy economy" usually refers to the use of pure hydrogen, the patterns in use of fuels over the past century, in which carbon-based fuels predominate, are nevertheless marked by increasing use of fuels containing more and more hydrogen. Use of pure hydrogen as a global energy resource is still minuscule, but energy released by breaking chemical bonds involving hydrogen as a component in other fuels has been increasing. As a result of the way life on earth has evolved, all living organisms are composed primarily of carbon. However, the carbon-based living world is increasingly turning to hydrogen as the source of the energy it uses.

As shown in Table 2, measured in terms of the energy content per pound of fuel, coal contains about one and a half times the energy of wood, and oil contains almost three times as much. Natural gas contains slightly more energy than oil. The most potent fuel of all, however, is hydrogen, which contains 61,000 Btu of energy per pound – three times as much as oil.

Table 2. Moving toward Hydrogen: More Powerful, Cleaner Fuels

Fuel Type	% Hydrogen	Energy Content (Btu per pound)	Particulates	Carbon Dioxide
Dry wood	5	6,900	5.22	775
Coal	50	10,000	5.00	240
Oil	67	19,000	0.18	162
Natural Gas	80	22,500	<0.01	117
Hydrogen	100	61,000	0.00	0

INFORM calculations based on energy content and conversion data from US Energy and Information Administration, *Annual Energy Review 1993;* and I. Ali and M. Basit, "Significance of Hydrogen Content in Fuel Combustion," *International Journal of Hydrogen Energy,* December 1993; and US EPA, *Compilation of Air Pollutant Emission Factors,* 1993.

Early steam locomotives burned wood, but this practice was abandoned as soon as coal came onto the market. Coal was tested as a fuel in early steam-powered automobiles, but it lost out quickly when oil, with its higher energy content, became available. Oil is now the primary transportation fuel, even in airplanes; however, in space applications, where fuel weight is a vital limiting factor, lightweight hydrogen has long been the fuel of choice. It costs about $4,500 in fuel and other expenses to lift each pound of the space shuttle from the ground into orbit; the less weight devoted to fuel, the more weight available for cargo, or the less expensive it is to propel the craft into orbit. Even using hydrogen, the space shuttle lifts off at each launch carrying 220,000 pounds of hydrogen.[25]

More Hydrogen: The Trend toward Cleaner Fuels

The choice of fossil fuels containing less carbon has lessened air pollution over the past century (per pound burned). Table 2 shows emissions of two key pollutants, carbon dioxide and particulate matter resulting from the uncontrolled burning of various fuels for which data are available. Carbon dioxide emissions from oil-burning, for example, are about one-third less than the carbon dioxide emissions from burning coal to release the same amount of energy.[26] Carbon dioxide emissions from natural gas-burning are about half those from coal; hydrogen use obviously emits no carbon dioxide.

At the same time, the world's increased demand for energy – in transportation and other applications – has led to an overall increase in air pollution during the fossil fuel era. The evolution in energy use suggests that the increased buildup of carbon dioxide in the atmosphere from fossil fuel burning may begin to ebb as a result of the patterns in energy use already underway. Even though atmospheric carbon dioxide emissions are still increasing as fossil fuel use continues, energy sources containing less and less carbon are gradually accounting for higher percentages of the world's energy supply. If this trend continues, carbon dioxide emissions will peak and begin to decline as natural gas and renewably derived hydrogen resources, coupled with many efficiency improvements, lessen demand for coal and oil.

Emissions of particulate matter – soot, smoke, and dust – from burning fossil fuels without air pollution control equipment declines in a similar manner, as is also shown in Table 2. Wood- and coal-burning are notorious sources of smoke, emitting 5 or more pounds of particulate matter during the release of 1 million Btu of energy. Particulate emissions from oil-burning drop dramatically, to just 0.18 pounds per million Btu of released energy.[27] Particulate emissions from natural gas-burning are minuscule and totally absent from hydrogen-burning.

Into the Era of Sustainable Energy

Sustainable, clean, nonfossil solar energy resources will most likely provide the world with most of its energy in the future. Developing the technologies to tap these resources is the critical issue in this third and final stage in the transition of world energy use from renewable energy, through the fossil fuel era, and back to renewable energy resources.

Moving into the sustainable energy era as quickly and easily as possible is one of the most important challenges facing the world today. The Marchetti analyses suggest that the trends in world energy use are already moving in the direction of some of the desirable attributes of a sustainable energy economy: cleaner and lighter fuels with decreasing amounts of carbon and increasing concentrations of hydrogen. Thus steps are already being taken toward a sustainable energy future, even as total fossil fuel use continues to climb.

The Marchetti depiction of fossil energy use indicates that historically it has taken about 50 years for a new fuel source to reach 50 percent of its maximum usage worldwide. It is unlikely that the transition to a sustainable energy economy will proceed much more quickly. A global shift to increased use of natural gas appears to be under way, but continuation of this trend is not guaranteed, nor is it likely that all regions will share equally in the expansion. Natural gas is not available in all markets, and where it exists it faces stiff competition from coal and oil. However, the quicker natural gas displaces coal and oil as the predominant fossil fuels, the less the environmental harm in the final stages of the fossil fuel era will be. Furthermore, the faster solar-derived hydrogen begins to displace fossil fuels, perhaps gaining its first toeholds by piggybacking into markets where natural gas has already made inroads, the sooner the benefits of a sustainable energy economy will begin to fully emerge.

Notes

1. "Individual transportation" refers to personal means of transit as opposed to group-based transit such as trains or buses.
2. An excellent review of the history of energy use, including food, other renewable resources, and fossil fuels, appears in C. and J. Steinhart, *Energy: Sources, Use, and Role in Human Affairs* (Duxbury Press, North Scituate, MA, 1974), pp. 33-126.
3. Population statistics from *Information Please Almanac 1995* (Houghton Mifflin, New York, 1994), p. 135. Energy statistics from Steinhart, *Energy: Sources, Use and Role in Human Affairs, op. cit.*, p. 33; and calculations cited in note 5.
4. United States Energy Information Administration, *International Energy Annual 1992* (Washington, DC, 1994), p. xvi; biofuels data added from London British Petroleum, *British Petroleum Statistical Review of World Energy* (British Petroleum, London, 1989), cited in *International Journal of Hydrogen Energy,* Vol. 18, No. 10, October 1993, p. 859.
5. Unless otherwise noted, statistics regarding world energy use are from United States Energy Information Administration, *International Energy Annual 1992, op. cit.*

6. Based on world population of 5.6 billion and daily food intake of 2,000 calories. One food calorie equals 4185 joules, or about 4 Btu, according to Diagram Group, *Comparisons* (St. Martin's Press, New York, 1980), pp. 126, 138, and 140.
7. R. Friberg, "A Photovoltaic Solar-Hydrogen Power Plant for Rural Electrification in India," *International Journal of Hydrogen Energy,* Vol. 18, No. 10, October 1993.
8. J. Ausubel, "Hydrogen and the Green Wave," *Proceedings of the First Annual Meeting of the National Hydrogen Association* (Washington, DC, March 8-9, 1990), pp. 20-13 and 20-23; and Daniel Yergin, *The Prize* (Touchstone Press, New York, 1991), p. 80.
9. Steinhart, *Energy: Sources, Use, and Role in Human Affairs, op. cit.,* p. 35.
10. J. Flink, *The Car Culture* (MIT Press, Cambridge, MA, 1975), pp. 34, 38.
11. J. Ausubel, "Hydrogen and the Green Wave," *op. cit.,* pp. 20-23.
12. Steinhart, *Energy: Sources, Use, and Role in Human Affairs, op. cit.,* p. 36.
13. Steinhart, *Energy: Sources, Use, and Role in Human Affairs, op. cit.,* pp. 91-125; and "petroleum" and "coal" listings in *Compton's Interactive Encyclopedia 1992* (CD-ROM version).
14. Unless otherwise noted, statistics about energy use in the United States were obtained from the United States Energy Information Administration, *Annual Energy Review 1993* (Washington, DC, 1994).
15. C. Marchetti, "How to Solve the Carbon Dioxide Problem Without Tears," *International Journal of Hydrogen Energy,* Vol. 14, No. 8, August 1989.
16. Two articles that discuss the history of fossil fuel use are T. Veziroğlu, "Hydrogen Technology for Energy Needs of Human Settlements," *International Journal of Hydrogen Energy,* Vol. 12, No. 2, February 1987; and F. Barbir *et al.,* "Modeling of Hydrogen Penetration in the Energy Market," *International Journal of Hydrogen Energy,* Vol. 18, No. 3, March 1993.
17. Steinhart, *Energy: Sources, Use, and Role in Human Affairs, op. cit.,* p. 60.
18. Barbir *et al.,* "Modeling of Hydrogen Penetration in the Energy Market," *op. cit.*

19. Articles about energy use throughout the fossil fuel era used in writing this section include: C. Marchetti, "When Will Hydrogen Come?" *International Journal of Hydrogen Energy,* Vol. 10, No. 4, April 1985; D. Scott, "Hydrogen in the Evolving Energy System," *International Journal of Hydrogen Energy,* Vol. 18, No. 3, March 1993; and H. Rogner, "Fuel Cells, Energy System Evolution and Electric Utilities," *International Journal of Hydrogen Energy,* Vol. 19, No. 10, October 1994.

20. D. Scott, "Hydrogen in the Evolving Energy System," *op. cit.*

21. K. DeJong and M. Van Wechem, "Carbon: Hydrogen Carrier or Disappearing Skeleton?" *International Journal of Hydrogen Energy,* Vol. 20, No. 6, June 1995; and C. Marchetti, "When Will Hydrogen Come?" *op. cit.*

22. D. Yergin, *The Prize, op. cit.,* p. 43.

23. D. Considine, *Energy Technology Handbook* (McGraw-Hill, New York, 1977), p. 3-175.

24. American Gas Association, *1993 Gas Facts* (Arlington, VA, 1994).

25. Presentations made by NASA officials during a field visit to the Kennedy Space Center, June 20, 1994.

26. United States Environmental Protection Agency, *Compilation of Air Pollutant Emission Factors* (Washington, DC, 1993).

27. *Ibid.*

Chapter 2

Findings and Conclusions

Twentieth century civilization is in the midst of an era of rapid change with regard to the options for personal transportation – how individual people move about in an increasingly mobile world. When INFORM set out a few years ago to investigate the potential for hydrogen to supply the growing demand for transportation energy, the first hydrogen fuel cell vehicle had not yet been demonstrated. Today, fuel cell vehicles are a "real world" option, although their technology still needs refinement and their viability in the market depends on the cost-reducing economics of mass production.

The few vehicles now equipped with fuel cells offer a glimpse of a radically different personal transportation future powered by sustainable and pollution-free energy. Along with other sustainable fuels and automotive technologies, such as electric cars powered by batteries, hydrogen vehicles offer a real alternative to the environmental and energy supply problems that plague, and will eventually undermine, the conventional automobile. Not since the gasoline-burning automobile replaced the horse as the kingpin of personal transportation a century ago has the prospect of fundamental change been as exciting and potentially rewarding as it is today.

If a transportation system powered by clean, renewable resources is our eventual goal, what technological and economic issues must be addressed to make that a reality? INFORM has identified viable pathways that can take us from the polluted present to a cleaner future – from our current near-total reliance on oil as a transportation fuel to a future in which we could transform the energy of the sun into pollution-free hydrogen vehicle fuel.

The principal barriers to the advancement of hydrogen vehicles are the dearth of political leadership, the absence of a national commitment to fundamental change, and the lack of adequate public- and private-sector financial support – not insurmountable technical obstacles or severe economic repercussions. A substantial but manageable research investment over a decade would likely lead to the development of a variety of vehicles fueled by hydrogen that perform as well or better than the gasoline vehicles of today, with a small fraction of the environmental impact.

The opportunities for innovation and economic growth in the field of hydrogen energy are largely untapped, and many nations are working to establish their position in this fledgling field. Today, Germany and Japan are leading the way in hydrogen vehicle technology. At the same time, there is a growing interest in multinational projects, most of which are proceeding without the participation of the United States. It remains to be seen what role the nations of the world will take in providing new automotive fuels and technologies. It is clear, however, that for the United States the implications of the emergence of hydrogen vehicle technology go far beyond the types of cars American consumers may buy in the future; the stakes include the competitiveness of American industry in the global transportation marketplace of the 21st century.

This chapter presents INFORM's major findings and provides conclusions from our research for policymakers and business, scientific, and environmental leaders to consider. The findings are grouped into eight categories:

- Global energy transitions
- Energy for transportation
- Hydrogen automotive engines
- Hydrogen storage
- Hydrogen production and distribution
- The hydrogen fuel cycle: efficiency, cost, environmental impact, and safety
- United States programs focusing on hydrogen as a transportation fuel
- International hydrogen vehicle programs.

Evaluating Global Energy Transitions

The world is in the middle of the second major energy transition in human history. The first transition occurred when fossil fuels (coal, oil, and natural gas) replaced renewable resources (food and wood) as the world's major sources of energy. The uncoupling of energy supply from biological processes has made possible a massive increase in energy use during the past century – an increase that has significantly contributed to the high standard of living enjoyed by industrialized countries in the 20th century. The second energy transition has been marked by a shift away from fossil fuels to renewable resources that humans were previously unable to harness. The challenge is to expand renewable energy use so that fossil fuels can be replaced without decreasing the supply of energy or diminishing the quality of human life.

A look at current patterns in energy use reveals that a transition to cleaner, more powerful fuels containing less carbon and more hydrogen has been underway for more than a century. The evolution in energy resources since the mid-1800s shows that there have been upswings and downswings in the use of specific fossil fuels over time: first coal, then oil, and, increasingly, natural gas. Each successive fuel contains less polluting carbon and more energy-dense hydrogen than its predecessor. Pure hydrogen contains more than three times as much energy per pound of fuel as oil.

The transition to a sustainable global energy system may begin first in the transportation sector because this is an area where the world's energy and environmental problems are particularly severe. Automotive exhaust emissions are the largest single source of air pollution in the world today, especially in urban areas, where they often contribute more than 60 percent of all air pollution. In the United States, about half of all air pollution regulated under federal law and 31 percent of the carbon dioxide emissions implicated in greenhouse warming comes from automobiles. Nearly all of the more than 500 million cars and trucks on the road worldwide burn gasoline or diesel fuel, both of which are refined from oil. Oil, the most limited and rapidly depleting fossil fuel, now supplies 35 percent of the world's total energy demand.

Hydrogen is the key to sustainable transportation because it can be produced in virtually unlimited quantities from renewable resources, and because its use is nearly pollution-free. Hydrogen is a chemical element that accounts for more than 80 percent of all the matter in the universe, but most of this hydrogen is part of compounds containing other elements. Today, more than 99 percent of the hydrogen produced in the world is obtained from fossil fuels, and virtually all of this hydrogen is used in the chemical industry as a feedstock. If this hydrogen were used as an energy resource, it would supply about 1 percent of the world's energy demand. Hydrogen can also be produced using renewable energy resources as part of a sustainable transportation system. For example, solar energy can be used to extract hydrogen from water (a chemical compound containing two hydrogen atoms and one oxygen atom). Once created, pure hydrogen can be burned in an internal combustion engine to power conventional automobiles or used in a fuel cell to power an electric vehicle. Either way, using hydrogen recreates water and reestablishes the source of additional hydrogen production, thereby completing a cycle that can be continued indefinitely. Thus, hydrogen produced from renewable resources may provide the long-term solution to the problems created by reliance on oil in transportation.

Replacing conventional gasoline and diesel-fueled vehicles with natural gas could ease the transition to vehicles powered by hydrogen – a gaseous fuel in many ways similar to natural gas. Natural gas already powers about 750,000 vehicles worldwide. It is the best immediate replacement for oil-derived vehicle fuels from the standpoint of environmental benefits, adequate supply, energy security, economic viability, and safety. As automotive fuels, natural gas and hydrogen are linked in several ways:

- Natural gas and hydrogen can both be burned in internal combustion engines.
- Hydrogen can be added to natural gas to make it burn more cleanly.
- Both fuels share similar automotive storage and refueling system technologies.
- Most manufactured hydrogen is currently extracted from natural gas.

- Some hydrogen could be distributed through existing natural gas pipelines, while the construction of new pipelines to carry hydrogen could benefit from the "rights of way" for natural gas distribution.

Promising synergies also exist between hydrogen fuel cell technology and electric vehicles powered by batteries. The electric motor, controller, and many of the other components for electric battery-powered vehicles are identical to the systems used in hydrogen fuel cell vehicles. Hybrid electric automobiles containing both hydrogen- and battery-based propulsion systems show great potential for application during the next decade.

Harnessing Energy for Transportation

At the turn of the last century, the first "revolution" in personal transportation – the development of the automobile – was largely over, yet horses were still the dominant mode of transportation. During the period from about 1860 to 1910, the oil-based fuel and internal combustion engine technologies that make up automotive transportation as we know it were first commercialized. But it was not until the early 1920s that public acceptance of the automobile and the mass production system to produce it resulted in the number of automobiles on the road – 20 million – surpassing the horse population of the turn of the century.

The current development of electric vehicles, including those powered by hydrogen engines and fuel cells, represents a second transportation revolution. Like the first transportation revolution a century ago, the basic fuel and engine technologies for an improved method of personal transportation are being perfected, but in 1995, as in 1895, they are not yet widespread commercially.

There are four interconnected reasons why alternative transportation fuels – those not derived from oil – will become the norm for the world's passenger vehicles by the middle of the 21st century:

- **Supply:** World oil reserves are rapidly diminishing.
- **Environment:** Pollution from vehicles is creating an atmosphere that is increasingly damaging to public health and to the environment.

- **Economics:** The costs of producing oil and regulating the by-products of oil consumption continue to increase.
- **Energy security:** The political consequences of maintaining energy security in international markets are becoming untenable.

Alternatives to oil may find markets in modernizing nations sooner than in the rest of the industrialized world, because these countries have little or no economic investment in a transportation system based on the conventional gasoline-powered vehicle, and they can benefit from choosing a cleaner option. In the United States, there is one vehicle (car, truck, or bus) per 1.7 people. In contrast, China, where there are 652 people for every vehicle, and other countries building personal transportation systems from the ground up can look to cleaner alternative fuels and engine technologies to avoid the problems associated with oil use.

The Promise of a Sustainable Transportation System

A sustainable energy resource is one that is available in quantities large enough to be essentially limitless and that can be tapped in a manner that produces so little pollution that it is essentially pollution-free. The promise of a sustainable transportation system relies on transforming renewable resources into clean transportation energy, called an "energy carrier," that can power vehicles. Hydrogen offers certain advantages over another potentially sustainable transportation energy carrier: electricity stored in an automotive battery. Refueling with hydrogen is quicker than recharging batteries, and hydrogen storage systems are much lighter and smaller than batteries.

Hydrogen produced by splitting water into hydrogen and oxygen using electricity generated by solar resources may be the optimum energy carrier for the sustainable transportation system of the next century. Tapping the sunshine falling on just 5 percent of the world's desert regions would supply enough energy to meet total world energy demand. Five forms of solar energy could be used to power the energy for hydrogen production:

- **Solar thermal power:** Using the heat from direct sunlight.
- **Photovoltaic cells:** Converting sunlight to electricity via photosensitive chemicals in the cells.
- **Wind power:** Tapping wind energy to turn electrical-generating windmills.
- **Hydropower:** Tapping the energy of falling water – the most widely used renewable energy resource today.
- **Biomass:** Burning plant matter to release energy, or converting it into a liquid or gaseous fuel.

Water is an ideal resource for sustainable hydrogen production. Hydrogen extracted from water and then burned or used in a fuel cell is re-released in the form of water vapor. The energy contained in the hydrogen in 1 gallon of water can propel a fuel-cell powered hydrogen vehicle as far as 1 gallon of gasoline propels a conventional vehicle. If all the world's vehicles were operating on hydrogen, less than one-tenth of 1 percent of Earth's fresh water reserves would be required to support them; all of this water would be replenished after use as part of a closed cycle.

The diverse forms of solar energy suffer from two major drawbacks that can be overcome using hydrogen as an energy carrier. Gathering a sizable amount of solar energy requires large land areas, often not readily available near major fuel markets. Solar resources are unequally distributed throughout the world and throughout the days and seasons of the year. Because it is easy to store and transport, solar-produced hydrogen can bring a continuous supply of energy to areas where land and solar resources are less available.

Comparing Different Types of Hydrogen Automotive Engines

Hydrogen has been demonstrated as a viable automotive fuel in three technological paths: Internal combustion engines, fuel cells, and electric hybrid propulsion. Internal combustion engines are connected mechanically to conventional vehicles, while fuel cells produce electricity to power electric vehicles. Hybrids may involve combinations of engines, fuel cells, and electrical storage systems. In all cases the sole by-product of using hydrogen is water.

Roughly 50 hydrogen-powered vehicles have been built in the last two decades, fewer than 20 of which are operating today. The modern era of the hydrogen powered combustion engine began in the late 1970s, when researchers in Germany, Japan, and the United States developed prototype automobiles. The world's first fuel cell vehicle – a bus – was unveiled in Canada in 1993, and the United States showed the first fuel cell car and the second bus within the following year. These hydrogen vehicles operate as part of demonstration programs.

Hydrogen Use in Internal Combustion Engines

Conventional combustion engines require modification, not major redesigning, to burn hydrogen. Fuel injection systems must be adapted to supply sufficient quantities of hydrogen, steps must be taken to avoid backfiring of hydrogen, and the ratio of air to fuel must be adjusted to obtain maximum efficiency and power. The proven, commercially available technology for using natural gas in combustion engines is similar to the technology needed to use hydrogen.

Burning hydrogen releases less pollution than burning any fossil fuel. Hydrogen combustion releases no carbon monoxide, hydrocarbons, or particulate pollution, although trace amounts of these pollutants are produced from the oil used to lubricate the engine. Hydrogen is also the only combustion fuel that releases no carbon dioxide, which accounts for about half of the atmospheric greenhouse gases that have been linked to global warming. The only major environmental challenge associated with hydrogen combustion is the emission of nitrogen oxides, formed when the heat from combustion causes nitrogen and oxygen in the air to fuse. Equipment and technology are available to control nitrogen oxide emissions from hydrogen-burning engines to very low levels.

Blending small amounts of hydrogen with natural gas reduces air pollution from natural gas vehicles and provides a practical way to introduce hydrogen as an automotive fuel. Several research projects in

Colorado, Pennsylvania, and Florida have tested the performance of dual-fuel vehicles, which typically burn mixtures of 5 percent hydrogen and 95 percent natural gas in an internal combustion engine. These tests show reductions in air pollutant emissions of 50 percent compared with the already low emission levels of natural gas vehicles.

Advantages and Challenges of Fuel Cells

Hydrogen fuel cells release even less pollution than hydrogen combustion engine vehicles; they are zero-emission vehicles. Because fuel cells require no lubricating oil, and because there is no combustion to generate the high temperatures that lead to the formation of nitrogen oxides, fuel cell-powered electric vehicles offer the cleanest way of using hydrogen.

Fuel cells are two to three times as energy efficient as combustion engines. In other words, the percent of energy contained in hydrogen that is used to propel a car is up to three times greater with a fuel cell than with an internal combustion engine, which loses most of the energy it generates as waste heat or friction.

Fuel cells and batteries offer alternative ways of delivering electricity to an electric vehicle. Batteries store electricity, previously generated by an outside source, in the form of chemical energy. In contrast, fuel cells actually produce electricity on board the vehicle: Hydrogen and oxygen merge in a fuel cell, forming water and releasing energy as electricity.

Of the five basic fuel cell technologies under development, proton-exchange membrane fuel cells currently offer the greatest potential for application in vehicles. Fuel cells were first developed as part of the United States' space program in the 1960s. Proton-exchange membrane fuel cells powered two of the world's first fuel cell vehicles in 1993. The other technologies include phosphoric acid, alkaline, molten carbonate, and solid oxide fuel cells.

Various technological hurdles must be overcome before fuel cells can effectively compete, in terms of overall performance and cost, with internal combustion engines in automotive applications. The eight key technological issues facing fuel cells are: weight; catalyst performance;

durability; contamination; temperature; water control; engine starts and variations; and fuel flexibility. Ongoing fuel cell demonstration projects worldwide will likely yield improved solutions to these technical challenges as researchers continue to refine these prototype vehicles.

Hydrogen Electric Hybrids

Electric hybrid vehicles, which combine onboard engines or fuel cells that generate power with electrical storage systems that store power (such as batteries), may offer greater market potential than vehicles powered solely by single systems. Demonstrations of hybrid technology involving hydrogen indicate that hybrid vehicles may be lighter, smaller, more versatile, and offer better performance than vehicles running solely on engines, fuel cells, or batteries.

Opportunities and Obstacles for Storing Hydrogen in Vehicles

Three critical fuel storage issues affecting the commercialization of hydrogen vehicles are the weight and volume of the fuel storage system and the speed or ease of refueling. These limitations can be minimized if hydrogen is used in fuel cells, because their inherently high efficiency reduces the amount of fuel that must be carried on board a vehicle.

The weight and volume of hydrogen fuel storage systems needed in fuel cell-powered vehicles are comparable to those of current systems used successfully in natural gas vehicles. However, the weight and space requirements for hydrogen used in a conventional internal combustion engine vehicle are three times greater, making the practicality of these vehicles less likely.

Hydrogen refueling technology in many cases can be adapted from existing technology for refueling natural gas vehicles. Thus, the emerging refueling infrastucture for natural gas vehicles can be modified in the future to form part of a hydrogen-based transportation infrastructure. "Reforming" natural gas into hydrogen at an existing natural gas vehicle refueling station may provide an attractive near-term strategy for producing and supplying hydrogen for transportation until a large hydrogen production and distribution system is in place. The materials and systems used to refuel natural gas vehicles are analogous to those used to refuel hydrogen-powered vehicles, and for hydrogen-natural gas mixtures, they are the same as for natural gas.

Borrowing "fast-fill" compressed-gas refueling technology from natural gas, and using it to fill hydrogen fuel cell vehicles, offers the fastest refueling time for a hydrogen vehicle: About 5 minutes, which is comparable to gasoline refueling today. Refueling a hydrogen vehicle powered by an internal combustion engine takes longer than refueling gasoline or natural gas vehicles because more fuel is needed.

In a liquefied form, hydrogen, like natural gas, offers greater energy density than in its gaseous form – reducing the space requirements of fuel storage by half, compared with compressed hydrogen gas. However, the extremely cold (cryogenic) temperatures needed to maintain these substances in liquid form present costly technical challenges for fuel production, storage, and handling.

There are other methods of storing hydrogen besides compression and liquefaction. In order of their current development and potential for application in vehicles, these technologies entail: storing chemical carriers that contain hydrogen; binding hydrogen to solids; and rusting sponge iron to release hydrogen. Of these technologies, only two are being demonstrated in vehicles: the use of methanol as a chemical carrier and hydride storage, in which hydrogen is bonded to a metal surface. Much broader laboratory research and testing is underway, however, covering the full range of potential hydrogen vehicle storage technologies.

For all three critical areas – weight, volume, and refueling speed – hydrogen fuel cell-powered electric vehicles offer advantages over battery-powered electric vehicles. Even the battery industry's best predictions for product development envision battery systems that are heavier, bulkier, and take considerably longer to refuel than fuel cell systems. Currently available lead-acid battery systems are more than 20 times heavier and four times larger (45 cubic feet versus 10 cubic feet) than compressed hydrogen systems offering the same driving range; if the battery industry's research objectives are met, lead-acid batteries would still be roughly 10 times heavier and three times larger.

Expanding the Hydrogen Production and Distribution Infrastructure

Most hydrogen produced today is made from natural gas in a process known as steam reforming. Because this thermochemical process is the cheapest and most firmly established method of producing hydrogen, it is likely to remain in the forefront until production technologies based on renewable energy resources become commercially competitive.

Fuel cell vehicles running on hydrogen produced from natural gas would use up less of the earth's reserves of natural gas than natural gas vehicles would use. The steam reforming process is up to 70 percent efficient, meaning that 70 percent of the energy in the natural gas is retained in the hydrogen. If this hydrogen were used in a fuel cell, which is at least twice as efficient as an internal combustion engine, the fuel cell vehicle would require less natural gas in its fuel cycle.

All of the methods of producing hydrogen from renewable resources face technical and economic issues that must be resolved if hydrogen is to fuel a sustainable transportation economy. Nonetheless, within a decade, applied research could allow each of the following methods to play a vital role in producing commercial quantities of hydrogen. The first two methods listed below use highly developed technology, while the latter three are still in the early stages of development.

- **Electrochemical processes:** Using electricity from renewable resources, such as hydropower, to split the water molecule through electrolysis.
- **Thermochemical conversion of biomass**: Converting organic matter into hydrogen through the gasification processes of partial oxidation, pyrolysis, and steam reforming.
- **Photochemical conversion**: Using sunlight to heat catalysts that trigger the splitting of water into hydrogen and oxygen.
- **Photobiological conversion**: Collecting hydrogen generated as a waste product by some strains of algae and bacteria.

- **Thermal decomposition of water:** Heating water to more than 5,600 degrees Fahrenheit to break the molecular bonds and make it decompose into its constituent parts.

The existing distribution system for hydrogen is extremely limited but technologically established; thus, the scale of the expansion necessary to serve the huge transportation energy market is the key obstacle to making hydrogen fuel widely available. There are only 450 miles of hydrogen pipelines in the United States today, compared with 1.3 million miles of natural gas pipelines and 200,000 miles of oil pipelines.

Because of the significant overlap between the technology and infrastructure to produce, store, and distribute both natural gas and hydrogen, the present and expanded use of natural gas as a vehicle fuel should facilitate the entry of hydrogen into the US energy market. Up to 20 percent hydrogen can be blended with natural gas and transported in natural gas pipelines – without any pipeline modification. Moreover, the experience gained by the transporters of liquefied natural gas would be an invaluable asset as the market for shipping liquefied hydrogen grows. Barges, tankers, and rail cars are currently used to deliver liquefied hydrogen to its main customer – the space program.

Modifying natural gas pipelines to carry pure hydrogen would require addressing three key differences between the two fuels: the lower energy density and weight of hydrogen; hydrogen's greater potential to leak; and the tendency for pipes to become brittle when exposed to hydrogen. While in theory these technical challenges could be overcome, there is almost no focused research on the subject of converting natural gas pipelines to carry hydrogen.

Examining the Hydrogen Vehicle Fuel Cycle

Different pathways of hydrogen production, distribution, and use as a transportation fuel bear with them different efficiencies, costs, environmental effects, and safety aspects: These pathways are called hydrogen fuel cycles. From a fuel cycle perspective, energy efficiency means the percentage of energy delivered to the wheels of a vehicle versus the total amount of energy expended during the entire process of fuel production, distribution, and use. Fuel cycle efficiency has a direct bearing on the economics, environmental impact, and safety risks presented by different cycles of hydrogen use in vehicles.

Efficiency

When used in a fuel cell vehicle, the fuel cycle energy efficiency of hydrogen can greatly surpasses the low efficiency – 12 to 15 percent – of gasoline vehicles. While burning hydrogen in a combustion engine provides even less energy efficiency than the gasoline-burning internal combustion engine, using hydrogen in electric fuel cell vehicles offers efficiencies of up to 30 percent when the hydrogen is produced from natural gas, or up to 20 percent if it is produced from sustainable solar photovoltaic technology. Fuel cell vehicle efficiencies that are twice those of conventional vehicles have been demonstrated, and efficiencies triple those of conventional vehicles are possible.

Hydrogen electric hybrid vehicles that combine batteries with redesigned internal combustion engines that burn hydrogen at a steady rate offer greater energy efficiencies – 20 to 25 percent – than hydrogen internal combustion engines alone. By comparison, electric vehicles equipped solely with batteries charged with electricity as an energy source produced from fossil fuels offer efficiencies of 20 to 29 percent.

Economics

The cost of hydrogen as a transportation fuel ultimately depends on five issues:

- **Fuel costs:** Hydrogen produced from natural gas currently costs only about 20¢ more per equivalent gallon than gasoline. Over the next decade, expanded production of hydrogen from renewable resources could yield prices per equivalent gallon of $1.45 to $6.90, depending on the resource. But fuel cells' greater energy efficiency could more than erase this difference at the pump, because only one-third to one-half as much hydrogen is needed to propel a fuel cell vehicle compared with the amount of gasoline needed to propel a conventional vehicle the same distance. Fuel costs actually drop by switching from gasoline to a fuel cycle based on hydrogen produced from natural gas and used in a fuel cell-powered vehicle.

- **Vehicle costs:** Technological refinement and mass production are needed to bring the cost of hydrogen vehicles to prices comparable to those of gasoline vehicles. For the next decade, hydrogen vehicles powered by internal combustion engines could cost only $1,300 to $5,800 more than a conventional gasoline vehicle. The price tag for fuel cell vehicles within a decade could be between $4,300 and $15,300 more than for a conventional car. As larger numbers of hydrogen vehicles are built, these price differentials will decrease.

- **Externalities:** Costs associated with automotive use that are not included in the price of owning and operating these vehicles are external costs. Using hydrogen as an automotive fuel will likely entail few or none of the externalities most widely identified with conventional gasoline vehicles: health costs from exposure to emissions – estimated to cost the nation $20 to $50 billion per year; the impact of carbon dioxide on global climate, estimated at $2.5 to $25 billion per year; worldwide environmental effects of oil drilling, refinery emissions, and oil spills – estimated at $2.5 to $9.0 billion per year; and the cost of maintaining a military presence in oil-producing regions abroad. In addition, other costs associated with gasoline use could also be eliminated. These include the cost of pollution control devices on new vehicles – $5 billion per year – the expense of operating emissions testing facilities – more than $500 million per year – and the expenditures of federal, state, and local pollution control agencies to develop and enforce vehicle pollution control standards.

- **Economic credits:** Government policies or the free marketplace could help equalize the price of hydrogen and gasoline by rewarding users of clean hydrogen vehicles economically. For example, the pollution reductions achievable with hydrogen vehicles could be worth several thousand dollars per year if vehicle owners were permitted to sell them as emission reduction credits to other parties required or desiring to reduce pollution.

- **Life-cycle cost:** Research indicates that fuel cell-powered vehicles, over a decade of use, will entail comparable or more favorable costs in comparison with gasoline vehicles – mostly due to the efficiency gains from using hydrogen in fuel cell engines and to taking advantage of natural gas as a low-cost feedstock for hydrogen production.

Environmental Impact

From production to distribution to end use, hydrogen fuel cycles are generally cleaner than the gasoline fuel cycle. Producing hydrogen from renewable resources and using it in a fuel cell vehicle eliminates all air pollution associated with the fuel cycle. The only major environmental consideration during the production and distribution of hydrogen from renewable resources is the large land area required by solar technologies. Producing hydrogen from natural gas, although not pollution-free, results in only a fraction of the environmental impact associated with the gasoline fuel cycle.

Safety

All fuels, including hydrogen, pose significant safety hazards, but the widely held public perception of hydrogen as an especially dangerous fuel is largely a myth. In reality, hydrogen has compiled an excellent safety record throughout decades of use in the space program. Contrary to popular belief, the fatalities in the infamous Hindenburg disaster were not caused by the dirigible's burning hydrogen. Of 97 passengers, 35 died – and they died from injuries suffered when they jumped from the dirigible or from burns caused by exposure to burning diesel fuel or cabin furnishings.

Hydrogen vehicles have not yet been widely used under "real-world" driving conditions. To ensure that they are handled in a safe

manner, certain characteristics of hydrogen must be addressed. Technological solutions to these safety challenges are either available or under development. Concerns about hydrogen's use as a fuel stem from its low ignition temperature; ability to burn in a wide range of concentrations with air; propensity to diffuse rapidly; and fast flame speed. Technological solutions to these safety challenges include strong odorant additives, additives for visibility and to reduce explosion risk, alarm detection systems, improved ventilation and containment systems, hydrogen-capturing canisters, embrittlement inhibitors, and new pipeline materials.

Summing Up United States Programs to Promote Hydrogen Use in Transportation

Interest in hydrogen as a fuel has grown dramatically in the United States since 1990, placing the country behind only Japan and Germany in terms of total annual public- and private-sector investment in hydrogen. The $10 million federal research and development program for hydrogen technologies, although still minuscule compared with other energy development efforts, has increased 10 times since the beginning of the decade. In 1995, spending by the United States government was surpassed only by that of the Japanese government ($23 million) and the German government ($12 million). At the same time, promising state, local, and private entrepreneurial hydrogen development efforts are underway.

Despite the increase in hydrogen research, the total effort by the federal government is still too small to catalyze a move toward a sustainable transportation system based on hydrogen. Annual federal expenditures for petroleum research are still 90 times greater than the national hydrogen budget. Hydrogen expenditures also pale in comparison with government investments in other innovative transportation technologies; in the past few years, the federal government has invested hundreds of millions of dollars in the development of batteries for use in electric vehicles – many times the amount spent on hydrogen fuel cells, which provide a viable and potentially more attractive alternative power source for electric vehicles. A greater commitment of federal dollars – already the largest source of funding for hydrogen research and development – is sound national energy policy and a step that is needed to spur greater private-sector involvement in hydrogen technology.

Achieving a sustainable transportation system requires coordinated federal leadership in establishing and directing a national commitment toward that goal. Sustainable transportation is an idea whose time has come, and hydrogen will play a part in it. A hydrogen-based transportation system addresses a multitude of public policy concerns – public health, global environmental protection, national energy security, and international economic competitiveness. Different federal agencies have principal responsibility for each of these policy areas – for example, the Environmental Protection Agency addresses automotive tailpipe emissions, the Department of Energy is concerned mainly with energy security, and the Department of Commerce focuses on economic development issues. Each of these agencies has some involvement in programs to promote hydrogen vehicle development, but these programs are not coordinated with the activities of other agencies and are often at odds with other initiatives, not only at other agencies but sometimes within one agency. The multidimensional nature of transportation energy issues, combined with the differing missions of individual government agencies, requires strong leadership to ensure that all federal programs are moving in concert toward a common goal. The space program enjoyed such leadership, coordination, and national commitment. To date, these are lacking with regard to sustainable transportation in general, and hydrogen vehicle development specifically.

Federal research and demonstration dollars are not being spent effectively. A redirection of funding priorities is needed to stimulate a competitive environment for the development of the new technologies needed to bring hydrogen vehicles into the marketplace. Demonstration of hydrogen vehicles in real-world conditions is the key to their successful introduction into the marketplace. Short-term research projects could be designed to advance and demonstrate hydrogen technologies so that the results of one project could be compared side-by-side with the results from other similar projects. For example, if grants of $1 million were given to support 50 projects – each aiming to design and build within two years a car powered by hydrogen – the cars could then be tested in a series of competitive road rallies held around the country and open to the public. This approach might be more effective in advancing hydrogen vehicle technology and in garnering public support for hydrogen than the combined efforts of several existing partnerships between federal government agencies and the major automotive manufacturers, although these partnerships are funded at a much higher level.

The major federal effort to develop hydrogen as an energy source is the National Hydrogen Program. This program, which had a $10 million budget in 1995, is managed by the Department of Energy's Office of Utility Technologies and includes many national laboratories as well as private academic and industry participants.

The major federal effort to develop hydrogen fuel cells is the National Fuel Cells in Transportation Program. This program, which had a budget of more than $20 million in 1995, is managed by the Department of Energy's Office of Transportation Technologies and includes as participants the United States Department of Transportation, the California South Coast Air Quality Management District, and various manufacturers.

Three other federal programs that do not focus primarily on hydrogen vehicle technology nonetheless include or could potentially include applications of hydrogen in transportation. These are the Electric Hybrid Vehicle Program, run by the Department of Energy; the Technology Reinvestment Program, directed by the Department of Defense; and the Partnership for a New Generation of Vehicle, coordinated by the Department of Commerce. Taken together, these joint public-private efforts account for much greater investment than either the National Hydrogen or the National Fuel Cell programs.

Among the states, overall efforts to encourage use of alternative transportation fuels, including hydrogen, are most ambitious in California. With the nation's strictest air quality standards and some of the worst air pollution in the world, California has numerous state, local, and private efforts to develop hydrogen vehicle technology. New York and Pennsylvania also lead in research on hydrogen storage, fuel cells, and dual-fuel vehicles.

On a municipal level, the governments of Denver and Los Angeles

have each supported advanced hydrogen vehicle demonstration programs. In 1989, Denver funded the world's first program to demonstrate that vehicles can run successfully on a fuel mixture of 95 percent natural gas and 5 percent hydrogen by weight. During the last few years, Los Angeles, mainly through the South Coast Air Quality Management District, has spent several million dollars to support a number of hydrogen demonstration projects within the Los Angeles basin and outside the region. The District currently supports hydrogen programs at a level second among government agencies only to the federal government. This includes a $1.7 million grant in support of the Department of Energy's fuel cell bus program.

Increased federal and state funding has led to rapid growth in private-sector hydrogen development activities. A 1994 National Hydrogen Association survey identified about 30 hydrogen research and development projects underway in the United States, nearly all of which were directed by private companies and academic institutions. At the same time, several trade and public interest organizations have recently appeared before Congress to endorse increased federal hydrogen energy activity, and several new trade organizations have begun to promote the development of hydrogen fuel cell technology. Yet private-sector hydrogen activity pales in comparison with the $37 billion the oil industry is spending to revamp its refineries to "reformulate" gasoline – making it burn slightly cleaner without addressing the nation's dependence on oil.

Looking at International Hydrogen Vehicle Programs

World leadership in hydrogen research, except for applications in space travel, has never been centered in the United States. In the 1990s, Japan's 28-year, $11 billion program eclipsed the German program as the world's most ambitious hydrogen development effort. Both countries have made it a national priority to develop hydrogen as an alternative to oil dependence.

The German government's involvement in hydrogen vehicle re-

search began in the 1920s, declined after World War II, and was rekindled in the wake of the OPEC oil embargo of 1973. Today's program includes vehicle production and testing, basic research into the technology for handling and storing hydrogen, and the development of renewable energy technology to produce hydrogen.

Europe's largest solar-hydrogen production facility is located in Germany. Opened in 1990, the facility uses 270 kilowatts from photovoltaic cells to power an electrolyzer that separates hydrogen from water. The facility, which also includes a liquefied hydrogen refueling station, is to be expanded. Project expenses are divided between German government agencies and the Solar-Wasserstoff-Bayern GmbH, a joint venture that acts as the project manager.

Japan's hydrogen program began in 1973 and was upgraded in 1993 as part of its "World Energy Network," which aims to develop and commercialize advanced energy technologies that reduce global carbon dioxide emissions. The Japanese program is the world's largest government-funded hydrogen development effort, with research concerning production, storage, and vehicle technology.

The leading auto manufacturers in designing prototype hydrogen vehicles are Daimler-Benz and BMW in Germany and Mazda in Japan. These companies' work includes the design and demonstration of hydrogen internal combustion engine and fuel cell vehicles, as well as research into liquefied hydrogen and metal hydride storage systems.

On a per-capita basis, Canada produces and consumes more hydrogen than any other nation; historically, most of this hydrogen has been produced from and used in its large fossil fuel industry. A major focus of the Canadian program has been to increase hydrogen production from its vast natural gas and hydropower resource base to serve world markets, while a reduction in oil dependence has been a secondary concern.

The local transit fleet in British Columbia, Canada, began testing the world's first fuel cell bus in 1993. Built by Ballard Power Systems, the bus runs on proton-exchange membrane fuel cells and can carry 20 passengers. A second and larger bus is now under development.

Several international hydrogen projects link energy producing

regions with countries that have high levels of energy consumption. The largest of these are:

- **The Euro-Quebec Project**: A joint undertaking of governments and industries that proposes to use electricity generated at a hydroelectric plant in Canada to produce hydrogen from water, then transport it for use in Europe. Included in this project are five hydrogen vehicle demonstration projects now underway in Europe and Canada.
- **The Hysolar Project:** A partnership between government and universities in Germany and Saudi Arabia that aims to produce hydrogen with solar photovoltaic cells and use this fuel in stationary and vehicle applications.

Conclusions

A fundamental change in personal transportation is not only inevitable but imminent. Today's gasoline-centered transportation fuel economy has created major problems in public health, the environment, and national energy security. Current government policies and technology development efforts designed to address these concerns through modification of conventional engines and oil-derived automotive fuels are facing diminishing returns. Any reasonable projection of expanded global transportation based on 20th century fuels and engines unravels in the face of environmental and oil supply constraints.

Hydrogen may be the ideal fuel for the vehicles of the future, providing the key to sustainable transportation. It can form the backbone of a sustainable transportation system because it can be made in virtually limitless quantities using renewable resources and, generating virtually no pollution, it can be used in either conventional vehicles or in fuel cells to generate electricity for electric vehicles.

There are no insurmountable technological obstacles to the development of commercially viable hydrogen-powered vehicles. The future of hydrogen as a vehicle fuel does not require as-yet-unknown technological breakthroughs or inventions. It depends, instead, on adequate public-

and private-sector investment and a national commitment to change.

The pace of recent advances in hydrogen vehicle technologies is startling, particularly those made in fuel cell technology. These developmental leaps rival those made in computers in the last two decades or in radio and television half a century ago. By any reasonable measure, the rate of technological progress in hydrogen research is faster than the progress in either conventional engine technologies or in developing batteries for automotive use.

Rapidly changing markets for transportation fuels and technologies may overtake today's automotive and fuel industries. A century ago, cars replaced horses. Today, zero-polluting electric cars powered by hydrogen offer advantages over conventional cars – advantages that stand in comparison with conventional vehicles today as the Model T vehicles stood in relation to the horses they replaced in the early 1900s. Stable owners represented society's transportation system 100 years ago, but they offered little to the engineers who designed the first automobiles. Similarly, today's major automobile manufacturers may be overtaken by developments in hydrogen-powered vehicles achieved by others – smaller companies and research laboratories.

There is an imbalance in America's energy research priorities with regard to hydrogen technologies. Far more money is being invested in attempts to achieve incremental improvements in conventional automotive and fuel technologies and in developing advanced batteries than is being spent for hydrogen technologies that show as much or more promise in solving our transportation problems. Stronger national leadership, increased funds for research and development, and programs designed to foster innovation by emerging private-sector industries are needed to solidify a national commitment to fundamental change in transportation through hydrogen vehicles.

Developing commercially viable and internationally competitive hydrogen vehicle technologies in the United States will require a greater commitment of federal dollars. Already, the federal government is the biggest source of funding for hydrogen research and development, yet government spending on hydrogen research and development lags behind spending by the governments of Japan and Germany.

Federally funded hydrogen research and development efforts lack the coordinated leadership that is needed to achieve optimum results. Putting together the various parts that will comprise commercially viable hydrogen vehicle technologies will require a level of coordination comparable to that of the space program. Instead, today's hydrogen research and development efforts are fragmented among federal agencies that fail to coordinate the various dimensions that hydrogen-based transportation addresses: public health, the environment, national energy security, and international economic competitiveness.

The future competitiveness of the American automotive industry requires investment in new technologies that will serve the needs of the fastest-growing international auto markets. With 19 million more vehicles than registered drivers, the vehicle market in the United States is nearly saturated. But the global appetite for cars has merely been whetted. China has one vehicle per 652 people; in India, people outnumber vehicles by 354 to 1; Africa's ratio is 71 to 1. Developing nations that want the benefits of modern transportation have not yet committed themselves to gasoline-powered vehicles. Rather than opting for infrastructures based on the 100-year-old American model, they are likely to choose cleaner, more abundant transportation fuels and technologies. To remain fully competitive, the United States will need to invest in the development of these fuels and technologies at least at a level comparable to the investments of its closest competitors.

One of the most important ways of encouraging the development of hydrogen vehicles is to support the widespread use of natural gas vehicles. Expanding the role of natural gas vehicles can ease the transition to vehicles powered by hydrogen. Both are gaseous fuels, and they share many characteristics. Hydrogen is a virtually pollution-free vehicle fuel; natural gas is the best immediate replacement for oil-derived vehicle fuels in terms of environmental performance, national energy security, economic viability, and safety. The development of an infrastructure for natural gas vehicles can pave the way for an infrastructure that will one day serve hydrogen-powered vehicles. Similarly, there are strong synergies between hydrogen technology and electric vehicle technology, so that electric vehicle research can be applied to hydrogen vehicles as well.

Chapter 3

Harnessing Energy for Transportation

In 1900, some 20 million horses formed the backbone of the personal transportation system in the United States.[1] There were more than 2,500 horses on the roads for every automobile.[2] At the same time, although it was barely noticeable, a transportation revolution was not only underway, it was nearly over. During a 50-year period, from about 1860 through 1910, the elements of automotive transportation as we know it today were developed and commercialized. The old personal transportation system, centrally and indelibly linked to human life for centuries, was swept aside permanently within a span of two generations.

During the course of this revolution a century ago, conventional transportation fuels, hay and oats, were replaced by liquid fuels refined from oil. This signaled a change from a sustainable transportation energy system based on renewable energy resources to a nonsustainable energy system based on a depletable resource. The conventional transportation engine, the horse's stomach, was replaced by the internal combustion engine. As these events unfolded, the public, in 1900 generally skeptical of the "horseless carriage," gradually grew not only to accept a change in personal transportation, but to embrace automotive technology with a remarkable and stubbornly persisting fervor.

The tale of the world's first transportation revolution reflects how rapidly change can occur when the needs and opportunities of the times change. It provides a backdrop, as well as lessons, for the second transportation revolution – already underway today.

The Transition to Gasoline

The revolution from horses to automobiles ultimately succeeded because the automobile proved itself to be a better product, but this took a while. The fuel element of this transportation revolution started first, followed by the development of the engine technology to use it. Underground reserves of oil were first tapped on August 29, 1859, when a Pennsylvania well hit oil. Oil was easy to handle and contained more energy per pound than either wood or coal, the predominant energy resources of that day. Almost overnight, kerosene refined from oil found a niche in the lamp market, providing more light for far less money than lamps burning whale oil. The fact that whales were being hunted to near-extinction in many of the world's oceans, raising prices and limiting supplies, contributed to this market opportunity. The combination of effects – growing problems in supplying whale oil and the emergence of oil as an abundant, superior, and low-cost fuel – catalyzed a change in fuel use. [3]

Oil gradually found other markets in home heating, powering industrial equipment, and as a fuel in some of the early automobiles; by the 1890s, John D. Rockefeller's Standard Oil was one of the most powerful corporations in the world. At one point in the late 19th century, Standard Oil controlled more than 90 percent of the oil-refining capacity in the United States.

World oil production is now more than 100 times greater than it was a century ago, but the fuels revolution was over by 1900. By that time oil had become a widely used commodity, the purchase and use of which played an essential part in most people's daily lives. Oil was so important, and Standard's grip on supplies and prices so tight, that the company became the target of a landmark antitrust action. In 1909, after 15 years of courtroom battles, Standard Oil was "busted" and the company dissolved into a number of smaller entities. What followed has been a century of incremental, though rapid, growth in oil production and use.

The introduction of gasoline as a transportation fuel to replace traditional oats and hay fed to horses marked one of the most elemental shifts in energy use in history. Oats and hay are renewable agricultural crops, which derive their energy from the sun as the plants grow. They are also solid fuels, like wood and coal – the other major fuels used at the time. Oil, on the other hand, is a nonrenewable resource, the fossil remains of plant life buried beneath the Earth's surface under great pressure for millions of years. Oil is also a liquid fuel, requiring different types of distribution and storage – barrels and pipelines – from that used for any of the solid fuels then in widespread use.

Development of Internal Combustion Engine Technology

The revolution in transportation engine technology started nearly two decades after the discovery of the fuel that was destined to power it. In 1877, Claus Otto patented the internal combustion engine in Germany. Light, compact, and powerful, the internal combustion engine was ideally suited for use in small personal vehicles. For example, a typical 100-horsepower automotive engine literally has the power of 100 or more horses, yet it weighs half as much as one horse and never needs a nap or a rubdown. The internal combustion engine also compared favorably to the bulky and noisy steam engines commonly used to power locomotives and most machinery invented during the industrial revolution. As with gasoline versus other fuels, the internal combustion engine in the late 19th and early 20th centuries quickly emerged as the best choice for a new method of personal transportation.

Although many fuels were tested in early internal combustion engines, the market was flush with cheap oil and it was not long before gasoline refined from oil emerged as the fuel of choice. Gasoline wasn't necessarily the best option – gaseous fuels are easier to ignite and are cleaner-burning than liquid fuels – but the high concentration of energy per unit of volume and the ease of handling a liquid compared with a gas contributed to gasoline's rise as a transportation fuel. Smelly and dirty as gasoline was, its use, especially in the small numbers of vehicles on the road at the turn of the century, did not offend the environmental sensitivities or laws of that era.

Gottlieb Daimler, a German, built the first automobile equipped with an internal combustion engine and powered by gasoline in 1885. He later teamed up with another German, Karl Benz, and in 1902 their company produced the world's first full-featured "driving machine," named after Mr. Benz's daughter, Mercedes. In 1899, the first of 2.7 million people in the United States to be killed in automobile accidents – H. H. Bliss – died when he stepped off a trolley on Central Park West in New York and into the path of an oncoming car.[4]

In 1900, motorists had registered a paltry 8,000 automobiles, only about 20 percent of which were powered by gasoline-burning internal combustion engines. Nearly 40 percent were electric vehicles, and another 40 percent were powered by steam boilers. These two automotive technologies enjoyed brief periods of popularity only to lose out to the more powerful, compact, and easy-to-fuel gasoline-powered engine.[5] It took until 1906 before automobiles powered by internal combustion engines surpassed those equipped with other types of engines.[6] By that year, cars accounted for 5 percent of the traffic in New York City, sufficient to trigger a public outcry that lead to the first posted speed limit, 8 miles per hour. This upper limit, only a pipe dream today on the city's crowded streets, was set to approximate the top speed of the transportation system's longtime standard-bearer, the horse-drawn carriage.[7]

In 1900, about 80 percent of the way through that 50-year transportation revolution, there was surprisingly little evidence from everyday life to suggest that a revolution was even underway. Horses still provided the dominant mode of personal transportation, and fundamental change, though imminent, was barely acknowledged and often greeted with skepticism by a wary public. As late as 1908, American automobile manufacturers produced a total of only 124,000 cars, compared with more than 2 million horse-drawn carriages built that same year. Yet the era of fundamental change for both fuels and engine technology for personal transportation was over because in 1908, Ford Motor Company manufactured the first Model T. Over the next 20 years, low-cost assembly line production implemented by Henry Ford manufactured 17 million Model T automobiles for an average price of less than $600 per vehicle. For the first time, cars became affordable to the public and they quickly relegated the traditional horse-based personal transportation system to the history books.

In 1912, the one-millionth automobile was manufactured in the United

States. By 1915, 2.5 million cars were registered in the country; only 2 percent were electric vehicles, the rest were powered by gasoline.[8] By 1925, the number of automobiles on the road finally surpassed the horse population of the turn of the century. Four years later, 26 million cars were registered in the United States. American auto production of 5.3 million cars in 1929 was 10 times higher than production in the rest of the world combined.[9] In this century, American manufacturers alone have built more than 500 million cars.[10] Like the expansion in oil production, automobile manufacturing has grown rapidly yet incrementally this century. The propulsion systems have not changed fundamentally: Internal combustion engines still power virtually every car.

As the 20th century draws to a close, the United States once again faces a day of reckoning in personal transportation. The total dependence on oil and the 19th century engine technologies used to burn it cannot sustain unbridled growth in automotive use. Continued growth in oil use is leading inevitably to resource depletion, preceded by eroding economics of production, increased pollution from production and use, and tragic social and political repercussions from efforts to secure supplies.

The signs of strain are already mounting. So are the signs that, like a century ago, better products are in the wings, capable of propelling the world through another transportation revolution. Like the previous transportation revolution, the modern-day revolution involves fundamental shifts in fuels, energy systems, and engine technologies. Like the earlier revolution, a great deal of work is underway, yet very few alternative vehicles have entered the marketplace. Also reminiscent of previous times, the attitude of the car-driving public today is skeptical of the prospects for a rapid transformation in the way people move about. Yet, as in 1900, much of the modern-day transportation revolution is already over.

The Age of Automobiles

Life in the industrialized world is characterized by a high level of convenience, comfort, and mobility – made possible in part by the personal automobile. Other nations are seeking to attain that "quality of life" associated with modern transportation systems. Automotive transportation already consumes more than one-third of the world's total oil production, and the global demand for automobiles has barely been touched. However, continuing to meet that demand with conventional technologies will occasion major environmental, economic, and political consequences.

In the United States today, there are 19 million more vehicles than registered drivers. (Unless otherwise specified, "vehicles" in this report refers to cars, trucks, and buses.) A new car is sold in this country every three seconds.[11] About 190 million registered vehicles are on US roads, comprising about 32 percent of the world's vehicles.[12] Gasoline or diesel fuel refined from crude oil powers virtually all vehicles today. Every second, American automobiles burn nearly 4,000 gallons of gasoline while traveling close to 100,000 miles. Per-capita energy use for transportation in the United States is three times higher than in Germany or France and five times greater than in Japan.[13] Every second, Americans send more than $1,500 overseas to pay for imported oil.

Worldwide, about 46 million new vehicles are now built each year, about 87 each minute, adding to the more than 595 million cars and trucks on the roads today.[14] The number of automobiles in the world has doubled approximately nine times this century. The latest doubling, from 300 million to nearly 600 million vehicles, has occurred in about 18 years. The doubling before that took about 12 years.[15] It appears clear that without a major change in the pattern of automotive manufacturing, the world vehicle population will likely surpass 1 billion around 2010 and more than double current levels before 2020.

Increased numbers of automobiles, of course, means more demand for oil until alternative fuels and technologies enter the market. Because the automobile as developed over the past century is only about 12 percent efficient in delivering the energy released from fuel-burning to the wheels (see Chapter 7), the conventional automobile represents one of the most wasteful, as well as one of the largest, uses of the world's most limited fossil fuel. The demand on oil is exacerbated by the fact that, as automotive technology and roadways have improved, each car is being driven more over its lifetime. The average person travels nearly twice as far by automobile now as in 1970.[16] American motorists drive each car 13 percent more miles annually than in 1970 and keep the average car on the road for 8.1 years, compared with 5.6 years.

Worldwide, there is one vehicle – car, truck, or bus – for every 11 people, compared with one vehicle per 1.7 people in the United States. In China, a country now experiencing extremely rapid economic growth, there are roughly 1.7 million cars for nearly 1.2 billion people, or one vehicle per 652 people. In India, people outnumber cars by 354 to 1; in Africa, the ratio is 71 to 1.[17] In underdeveloped countries, there is essentially no transportation infrastructure in support of the conventional gasoline-powered vehicle. How will these countries respond to the demand for personal transportation that will emerge as development occurs in the 21st century? Will they build a system modeled on the path pursued by the United States 100 years ago, when oil was plentiful and the internal combustion engine represented the state of the art? Possibly for the short term. However, it is likely that, starting with a clean slate, many modernizing nations will increasingly look to alternative transportation fuels and engine technologies to avoid the very problems that have emerged from oil use this century. These alternatives offer an ever-greater promise to meet the global demand for transportation far into the future. A commitment to oil-based transportation on the part of the United States may lead to a loss of competitiveness in new transportation fuels and technologies, especially in emerging global markets. As discussed in Chapter 9, other countries, notably Japan, Germany, and Canada, are moving aggressively ahead in developing hydrogen transportation technologies. The United States only stands to lose by not expanding its efforts to match the efforts of its international competitors.

The End of the Oil Era

It seems inevitable that alternative transportation fuels – those not derived from oil – will become the norm for the world's passenger vehicles by the middle of the 21st century. This move to alternative fuels is likely to take place for four interconnected reasons:

- World oil reserves are rapidly being depleted.

- Pollution from vehicles is increasingly damaging to public health while threatening to induce significant global climate change and other serious environmental effects. Efforts to reduce pollution are facing diminishing economic and environmental returns.
- The costs of producing oil and regulating the by-products of oil use continue to increase, while the cost of some alternative fuels is dropping.
- The political consequences of maintaining energy security in international oil markets are increasingly untenable.

Taken together, these four reasons underlie the urgent need to develop alternative transportation systems.

Limited Supplies

During every 20-year period this century, the world has consumed more oil than had been consumed in all previous history (see Figure 1). By the year 2000, total world oil use since its discovery in 1859 will exceed 800 billion barrels. Oil use during just the 20 years from 2001 and 2020 is likely to reach more than 1 trillion barrels, exceeding all the oil consumed during the 20th century and surpassing all the oil reserves known to exist in the world today.[18]

As a result of a century of exploration, the world's most readily exploitable oil reserves have been identified. New oil deposits will certainly be found, but increasingly this quest will be met by diminishing returns. This situation has already been encountered in the United States, the most extensively explored region of the globe. About 80 percent of the 2.9 million oil wells drilled worldwide through 1986 were located in the United States.[19] For many decades, the country led the world in oil production, largely due to its aggressive efforts to find and exploit oil reserves. As shown in Figure 2, in 1925, for example, oil fields in the United States accounted for 71 percent of world production. Now, the country's easily exploitable oil reserves have been exhausted, exploration elsewhere has uncovered new oil deposits, and the dominance of the American oil industry has faded. By 1945, the percentage of world oil production accounted

Figure 1. History of World Oil Use – Impact of Exponential Growth

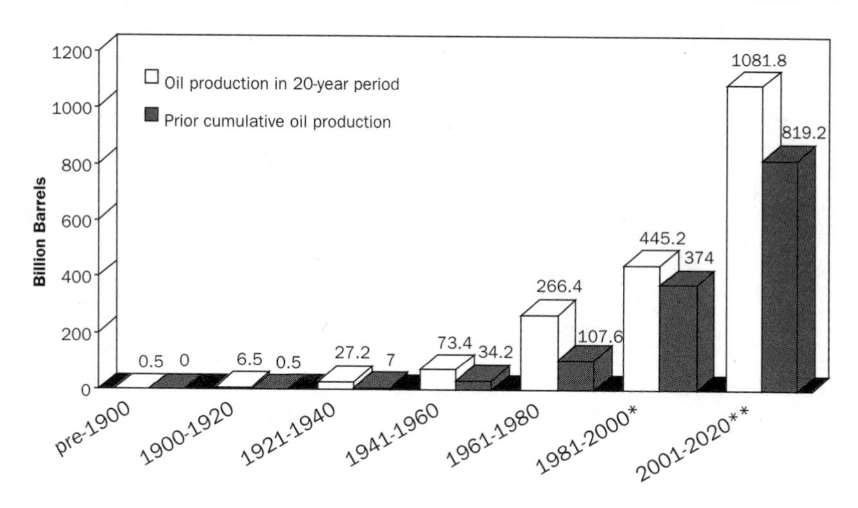

**estimated **projected*

Source: INFORM calculations based on data from D. Yergin, *The Prize;* and United States Energy Information Administration, *International Energy Annual 1992.*

for by the United States had dropped to 66 percent, and by 1965 it had fallen to 32 percent. American oil production now provides a paltry 12 percent of the world's total output. Even this level of production might be hard to sustain: The country has only 2 percent of the world's known oil reserves. On the other hand, it accounts for 36 percent of world oil consumption.

Depletion of oil reserves that can be inexpensively and easily tapped is already a well-established phenomenon in the United States. Domestic oil production peaked in 1970 and has dropped 20 percent since then, including declines in 8 of the last 9 years.[20] In 1986, the United States reached an important and unenviable milestone, precipitating a near-collapse in international oil prices: For the first time, oil use in transportation alone exceeded domestic oil production (see Figure 3). Transportation in the United States now consumes 30 percent more oil than is produced domestically.[21]

Figure 2. History of United States Oil Production – Percent of World Output

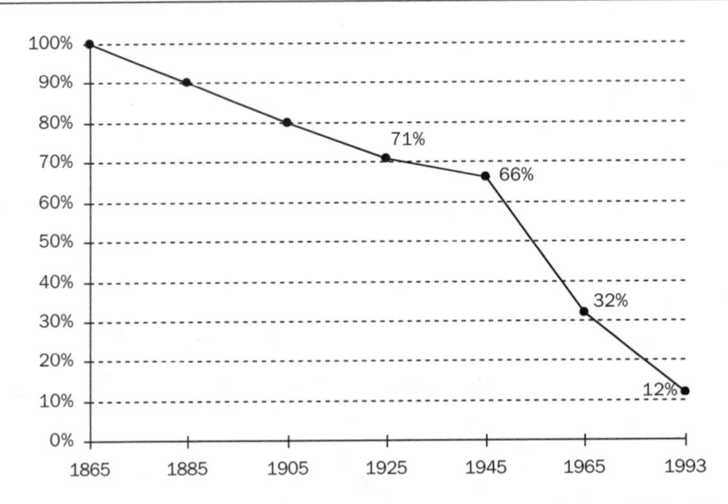

Source: INFORM calculations based on data from D. Yergin, *The Prize;* and United States Energy Information Administration, *International Energy Annual 1992.*

Figure 3. Oil Production, Transportation Oil Use, and Other Oil Use in the United States

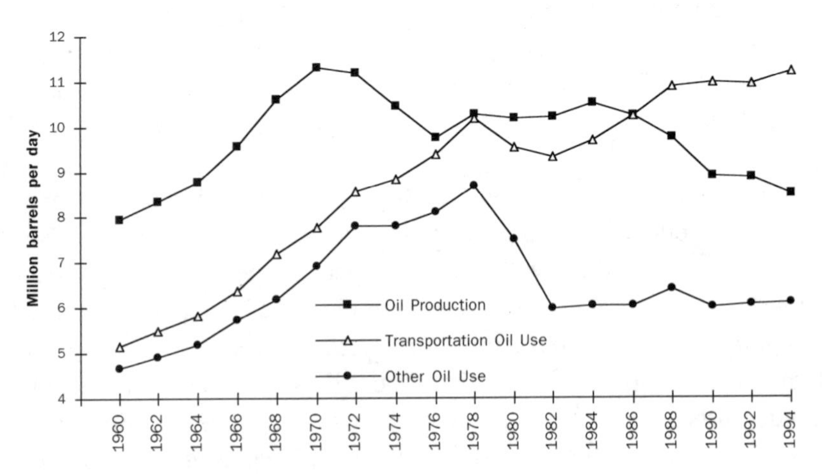

Source: United States Energy Information Administration, *Annual Energy Review 1993;* 1994 estimates by INFORM.

The decline in domestic oil production has exacted a heavy toll in jobs. According to the United States Department of Energy, about half of all oil industry employment, nearly 500,000 jobs, has been eliminated since 1982.[22] The United States Energy Information Administration expects this situation to deteriorate further. Under its "reference case" scenario (based largely on current trends), the EIA predicts that by 2010 oil use in transportation will grow by 2.6 million barrels per day and account for 80 percent of the projected increase in oil consumption for all uses. Domestic oil production, on the other hand, is forecast to drop another 1.85 million barrels a day.[23]

In 1992, four of the largest oil companies operating in the United States spent nearly twice as much money, $6.5 billion, to explore for oil in foreign countries than the $3.3 billion they invested in oil exploration at home.[24] The domestic investment and employment trends for oil run counter to the increased domestic investment and employment in renewable energy technologies and the environmental protection industries. A minuscule industry a couple of decades ago, environmental protection accounted for 4 million jobs in 1992 and represented a $100 billion-per-year business. Although still nascent, renewable energy industries have doubled employment several times over since 1970.[25]

Until alternative fuels come into widespread use for transportation, as they have in other energy sectors, the United States will have no choice but to turn to oil imports to meet its transportation energy demand. Formidable as this challenge may seem, it has in fact been successfully met in other oil uses. As shown in Figure 3, oil use in nontransportation applications – residential, commercial, and industrial uses and electrical generation – dropped precipitously in the late 1970s and have remained level since then, due mainly to increased energy efficiency and use of alternative fuels in these sectors. Although oil use in transportation dropped slightly, its upward climb was renewed in the early 1980s.[26]

The decline in oil production in the United States will be repeated in other countries as reserves are depleted. A 1990 report by the United States Geological Survey predicts that world oil production outside of the Middle East will decline after the turn of the century.[27] Opinions vary about future trends in global oil production, including the Middle East, but many, if not most, energy analysts agree that worldwide production is likely to peak sometime around 2010.[28] Unless changes in the patterns of oil use begin to happen soon, the 1993 report of the World Energy Council notes, "the realities suggest that the international energy scene, and many national energy situations will be even more difficult in 2020 than they were in 1990."[29]

Environmental Effects

The modern era of environmental protection, marked by the first "Earth Day" in 1970, has required an enormous industry effort to reduce pollution from all sources. Environmental laws have affected the automobile manufacturers as much as any other industry and more than most. According to the Department of Commerce, pollution abatement and control expenditures for automobiles in 1990 were about $81.8 billion.[30]

Automobiles emit four major pollutants that are regulated under the Clean Air Act of 1970 because they threaten public health and the environment:

- Carbon monoxide
- Nitrogen oxides
- Hydrocarbons
- Particulates (soot, smoke).

Health and Environmental Impacts of Automotive Exhaust

Health effects of exposure to carbon monoxide include impaired vision, memory, muscle coordination, and concentration; prolonged exposure may damage the heart and high exposure levels can cause death. Nitrogen oxides reduce resistance to respiratory ailments; they also contribute to acid rain and visible air pollution. Nitrogen oxides combine with hydrocarbons to form ozone – the cause of "smog" – which damages the eyes, skin, and lungs and also damages plants and buildings. Hydrocarbons themselves – more than 100 appear in automotive exhaust – include such toxic substances as formaldehyde and the known carcinogen benzene. Particulate matter impairs breathing, aggravates heart ailments, and reduces the body's defenses against respiratory diseases and cancer; it also damages crops, corrodes structures, and reduces visibility. Carbon dioxide has been implicated as a cause of global warming.

Figure 4. History of Automotive Emissions

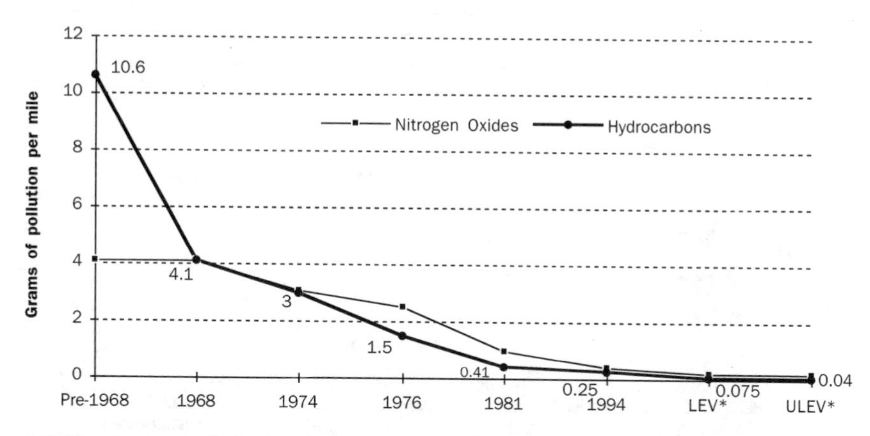

* *California standards for Low-Emission Vehicles and Ultra-Low-Emission Vehicles*

Source: United States Energy Information Administration, *Annual Energy Review 1993;* 1994 estimates by INFORM.

In addition, two other categories of air pollution, not currently regulated as individual pollutants, are emitted in large quantities in automotive exhaust: toxic chemicals, such as the known carcinogen benzene, and greenhouse gases, such as carbon dioxide.

Due to a long series of increasingly strict pollution control requirements, tailpipe emissions from a new automobile now contain 96 percent less carbon monoxide and hydrocarbon air pollution and 76 percent less nitrogen oxide air pollution than cars produced in the 1960s, before the Clean Air Act was enacted.[31] Additional control requirements that are beginning to take effect as a result of amendments to the Clean Air Act in 1990, and stricter standards enacted by the California Air Resources Board, will reduce automotive air pollution even further.

However, efforts to control pollution from conventional gasoline-burning, internal combustion engine-powered vehicles are now facing diminishing returns. Each round of pollution reductions has improved air quality in the United States, but these gains are quickly offset by the increased number of automobiles and the longer distances each vehicle travels during its life. As shown in Figure 4, there is little room left to reduce dis-

Figure 5. Air Pollution from Transportation, 1993: Pollutants Regulated Under the Clean Air Act

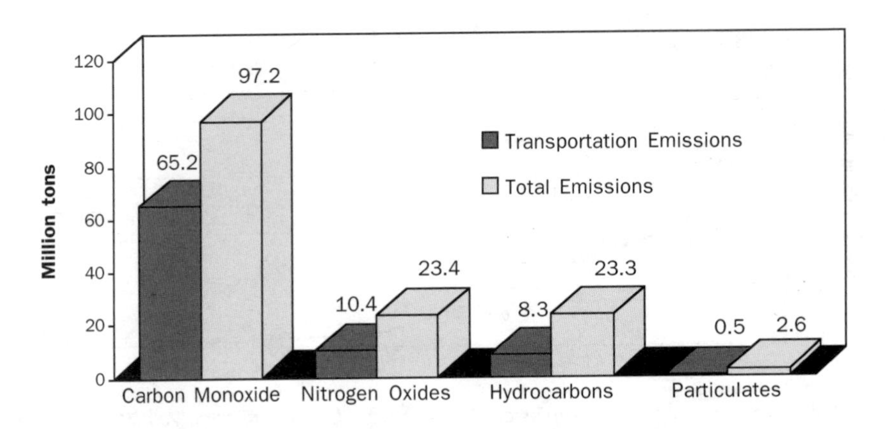

Air Pollutant	**Percent Emissions from Transportation**
Carbon Monoxide	67%
Nitrogen Oxides	44%
Hydrocarbons	36%
Particulates	19%

Source: US Environmental Protection Agency, *National Air Quality and Emissions Trends 1993.*

charges of pollutants from individual automotive emissions .

Vehicle emissions remain a major threat to public health and the environment. Nearly half of all air pollution emitted in the United States and regulated under the Clean Air Act is emitted by automobiles and trucks. As shown in Figure 5, in 1993 transportation accounted for 67 percent of the carbon monoxide emissions from all sources, 44 percent of the nitrogen oxide discharges, and 36 percent of the hydrocarbon pollution. Moreover, transportation accounted for about 19 percent of all particulate pollution in the US in 1993, about 80-85 percent of the toxic air pollutants emitted into the atmosphere, and about 31 percent of the carbon dioxide.[32]

The contribution of transportation air pollution to total air quality degradation is at its highest in densely populated urban areas, where it may account for more than 70 percent of total air pollution. The American Lung

Association estimates that the health costs associated with illnesses and premature deaths resulting from breathing automotive pollutants may total $50 billion annually.[33] A 1993 study by the Harvard School of Public Health concluded that between 50,000 and 60,000 people may die each year from exposure to just one pollutant, fine particulate matter, emitted primarily from trucks, buses, power plants, and factories.[34]

Carbon dioxide and global warming

Oil use is also the main source of increased emissions of carbon dioxide, which accounts for about half of the atmospheric greenhouse gases that may be causing a gradual but potentially catastrophic warming of the earth's atmosphere. Of the more than 6 billion tons of carbon that worldwide fuel-burning adds annually to the atmosphere in the form of carbon dioxide, about 17 percent stems from transportation energy use; oil-burning represents nearly all of this energy.[35] In the United States, automobile use of oil accounts for a much higher share of carbon emissions: 31 percent, or 437 million tons of the nation's 1.4 billion tons of carbon emissions in the form of carbon dioxide.[36] Every gallon of gasoline burned releases 20 pounds of carbon dioxide. American auto tailpipes emit 40 tons of carbon dioxide into the atmosphere every second.

The annual growth rate of carbon dioxide emissions in developing countries – where resource management and pollution controls are often lacking – is estimated to be three times the rate of increase in the industrialized world: 3.7 percent versus 1.2 percent.[37] Overall, the International Energy Agency forecasts that global carbon dioxide emissions will increase by 50 percent between 1990 and 2005 unless actions are taken specifically to reduce discharges.

Since the advent of the fossil energy economy in the middle of the nineteenth century, the concentration of carbon dioxide in the atmosphere has steadily increased from 270 parts per million to 350 parts per million – a 30 percent increase. Carbon dioxide levels are currently their highest in the last 160,000 years (although carbon dioxide still accounts for just 0.035 percent of all the gases in the atmosphere). Increased levels of carbon dioxide have been linked to an increase in Earth's temperature, called "global warming." During this century, the average global temperature has increased about 1 degree Fahrenheit. The six hottest years in this century all occurred in the 1980s. If fossil fuel use grows over the next 100 years as fast as during the past century, carbon dioxide levels in the atmosphere would more than double compared with today's already worrisome levels. Predictions of future temperature increases during the next 100 years range from 3 to 9 additional degrees, depending mostly on assumptions of future fossil fuel use.[38]

Many global climate experts fear that increasing carbon dioxide concentrations, in addition to rising temperatures, may also engender more frequent and more violent weather extremes. Evidence offered to support this theory includes the increase in weather-related disasters in the United States in recent years, including Hurricane Andrew in the South, the worst floods on record in the Midwest, and a number of droughts and fires in California. In the 20 years prior to 1987, the insurance industry paid no claims over $1 billion resulting from weather-caused disasters. Since then, there have been close to 20 natural catastrophes costing more than $1 billion each in claims.[39] Although not definitively connected to global warming, frequent, extreme swings in weather conditions may be an indirect effect of an overall warming trend.

Concern over global warming resulted in a meeting of international representatives in Rio de Janeiro for the Earth Summit of 1992. At the summit, more than 140 countries, including the United States, signed the United Nations Framework Convention on Climate Change. This document states its aims as stabilizing greenhouse emissions at a level needed to "prevent dangerous anthropogenic interference with the climate system." Many of the countries signing the Rio accord have established the actual rate of carbon dioxide emissions in 1990 as the stabilization target.

In 1993, Washington developed a plan to take the first steps toward implementation of the Rio accord. The Climate Change Action Plan, which aims to find a way of holding carbon dioxide emissions in the United States at 1990 levels by the year 2000, includes actions to reduce vehicle use through transportation control measures such as high parking fees.[40] Two years after adoption of the plan, carbon dioxide emissions are still far above the targeted levels. Even if successful, however, maintaining 1990 levels of carbon dioxide from transportation beyond 2000 will, by necessity, require measures to replace oil with fuels containing less carbon, or no carbon at all.

The Eroding Economics of Oil Use

The economics of oil have steadily deteriorated over the last 30 years as costs of oil exploration and production, as well as environmental requirements affecting oil production and use, have increased. The rising cost of producing oil from deeper, more remote, and less productive deposits represents the major cause of declining oil production in the United States. As easily exploitable fossil fuel reserves are exhausted, drilling and mining costs of the remaining deposits increase. The United States Energy Information Agency reports that the price of oil sold in the country increased 348 percent from 1970 through 1993. Adjusting for the effect of inflation still yields a 27 percent jump.[41] By 2010, the EIA predicts that the price of gasoline, adjusted for inflation, will increase an additional 15 to 50 percent, depending on the assumptions made in its analysis.[42] Although gasoline prices in the United States remain strikingly cheap, the cost of oil has, in fact, increased markedly over the past quarter-century and is likely to continue to do so in the near future.

Contrary to the trend in oil prices, the cost of using renewable energy resources has dropped dramatically during the past 20 years. In certain niche markets, solar energy today is cost-effective compared with fossil fuel use. The price per peak watt capacity of solar cells has fallen from $500 in 1970 to $4 today. Although still more expensive than electricity generated from fossil fuels, solar cells are now commonplace in watches, hand calculators, and several other markets. During the 1980s, the cost of electricity produced from wind technology dropped from 25¢ per kilowatt-hour to less than 10¢. Wind power is now comparable to the cost of burning coal to generate electricity in high-wind resource areas. If these price trends continue, within a decade or so the use of renewable energy will be cheaper than fossil fuels in many more, and perhaps in most, energy markets.

Domestically produced oil, although still cheaper than renewable energy in most uses, simply cannot match the price of cheaper oil from abroad. Consequently, the flow of imported oil to the United States has steadily grown. The United States became a net importer of oil in 1948, meaning it imported more oil than it exported. In 1973, the year of the OPEC oil embargo, imports accounted for 35 percent of America's oil usage. A national effort to reduce oil dependence in the aftermath of the embargo succeeded in cutting net imports to a post-1950 low of 27 percent in 1985, after which imports have again steadily risen.[43]

In 1994, for the first time in history, the United States imported more than half of all the oil it consumed. The 8.9 million barrels of oil imports daily comprised 50.4 percent of all oil consumed. Oil imports cost more than $50 billion in 1994, nearly half the nation's $108 billion total trade deficit. Meanwhile, domestic oil production dropped to its lowest level in 40 years.[44] The total cost of imported oil since the OPEC oil embargo 20 years ago, more than $1.2 trillion, is roughly equal to the cumulative US balance of trade deficit. According to the Energy Information Agency's "reference" case, oil imports will grow to 58 percent of all oil used in the country by 2010.[45]

An elaborate web of federal government subsidies has been established to keep the United States afloat in oil. Preferential tax treatment encourages domestic oil production despite the exhaustion of economically competitive oil reserves, and billions of military dollars are spent to keep the international flow of oil secure. Many recent studies have tried to quantify these subsidies. In early 1995, the Domestic Fuels Alliance, a trade group, reviewed seven such studies and found estimates of annual federal energy subsidies ranging from $10 billion, according to the Energy Information Administration, to $300 billion, calculated by Worldwatch Institute. Most studies, including a 1992 Congressional Research Service report, placed the subsidy around $25 billion per year.[46] These and other forms of "externalities," or external costs, are discussed further in Chapter 7.

The many environmental programs now affecting oil production and use are an additional factor in the eroding economics of oil dependency. Each automobile, for example, carries between $400 and $800 worth of gear to scrub its exhaust pipe fumes.[47] New refueling stations must be equipped with sturdier and more costly underground storage tanks to reduce the risk of groundwater contamination from leaks. Oil tankers are becoming more expensive to build and operate as national and international governments enact additional requirements to prevent spills in the wake of accidents such as the Exxon Valdez. Reformulated gasoline, with reduced concentrations of toxic chemicals, adds several cents to a gallon of fuel in the 10 heavily polluted major metropolitan areas where its sale has been mandated since 1995.[48] Yet environmental problems from oil use persist and will undoubtedly require new, ever costlier pollution-prevention measures in the future.

Not only are required pollution controls increasing the cost of operating gasoline-burning automobiles, but the economic costs of the environmental damage of the remaining uncontrolled pollution, termed "environmental externalities," are becoming more evident. Exposure to air pollution in the United States results in medical bills and lost worker productivity. Home values are lower in polluted areas. Pollution increases corrosion of buildings and bridges, reduces agricultural output, and decreases visibility – a form of aesthetic pollution. These are all examples of environmental externalities that are not reflected in the price of gasoline and diesel fuel, but which nevertheless are economic consequences of oil use.

Several states have attempted to quantify the price tag of environmental externalities and force polluters to incorporate them into the cost of their products. Massachusetts, for example, has set the environmental cost of one ton of nitrogen oxide air pollution at $7,934, while the global warming impact from carbon dioxide has been assessed a cost of $26 per ton.[49] If the cost of all the adverse effects of oil use in transportation were incorporated into the price of gasoline, it could add several dollars to the price per gallon, more than doubling the wholesale price today.[50]

Social and Political Repercussions

America's growing dependence on foreign energy resources raises the fourth reason for the end of the fossil fuel era. The adverse political and social consequences of securing continuous energy supplies, mainly oil, from abroad are increasingly problematic. The political climate in many of the world's richest energy regions is unstable and often outright hostile to the United States. Two-thirds of the world's proven oil reserves, for example, and 31 percent of its natural gas reserves are located in the Middle East.[51]

Securing access to these energy supplies has become a major focus of American foreign policy, and the overall success of these policies is double-edged at best and extremely expensive by any measure. There is little question that oil dependence was a motivating force in the Persian Gulf War, the staging of which involved 400,000 American soldiers and an expenditure of more than $50 billion. The cost of the Desert Storm invasion of Kuwait, if allocated differently, would have funded America's solar energy research and development program at 1990 levels for 1,437 years. As many people died during Desert Storm as were killed outright by the atomic bomb blasts at Hiroshima and Nagasaki combined, including about 150,000 Iraqi soldiers and citizens killed during the invasion.[52] Through this effort, the United States regained access to the petroleum supplies of Kuwait, a country that supplies less than 5 percent of the oil it consumes.[53] Various studies place the annual expenditures to maintain a military presence in the Middle East at between $0.5 billion and $50 billion, which translates to 1.5¢ to 30¢ per gallon of gasoline.[54] Other repercussions of dependence on foreign oil include the effects of limited competition due to a small number of participants in the oil market, as well as the macroeconomic costs to the US economy from adjusting to rapid, externally induced fluctuations in oil prices.

The Move to Alternative Fuels

During the past decade, the economic, supply, and environmental issues associated with oil-derived fuels have led to a change in the energy policies of many nations throughout the world, including the United States. These new policies, and the programs to implement them, herald the beginning of a global transition away from oil as the dominant transportation fuel and toward the use of cleaner, more abundant, and eventually sustainable energy resources – with hydrogen being the most promising long-term option. Although the number of alternative fuel vehicles is extremely small compared with conventional vehicles, their ranks are growing.

Most current alternative transportation fuel programs aim to increase use of the many fuels that are readily available to substitute for gasoline or diesel in conventional engine designs, while significantly reducing automotive air pollution. These alternative transportation fuels include natural gas, alcohol fuels such as methanol and ethanol, and liquefied petroleum gas. Liquefied petroleum gas (a by-product of natural gas production and oil refining) and ethanol (produced from corn, sugar, or other crops) each power about 3 million vehicles worldwide. There are about 750,000 natural gas vehicles. Fewer than 15,000 vehicles operating worldwide are powered by either methanol produced from natural gas or by electricity stored in batteries. Despite the inroads made by alternatives, however, gasoline and its oil-derived sibling, diesel fuel, still power about 99 percent of the world's vehicles.[55]

Federal and State Legislation

A number of federal and state government laws have propelled the growth in alternative transportation fuel use in the United States by promoting, and in many instances requiring, the use of alternative transportation fuels. Government-sponsored alternative transportation fuel programs now include: a broad array of research, development, and demonstration initiatives; firm schedules for the mandatory conversion of certain vehicle fleets to non-oil-derived fuels; tax and other economic incentives; and clear standards and regulations favorable to alternative fuel use.

Congress has enacted two laws that include comprehensive requirements for increased use of clean-burning alternative fuels in the transportation sector:

- The Clean Air Act Amendments of 1990 mandate the use of clean fuels in specified fleet vehicles in 22 highly polluted cities, home to 31 percent of the United States population. Beginning in 1998, 30 percent of new fleet vehicles in these cities must be capable of burning clean fuels. By the year 2000, 90 percent of vehicles in designated fleets must be powered by clean fuels. In a separate provision, the law requires that 300,000 alternative fuel vehicles be manufactured for sale in California in the 1999 model year.[56]

- The National Energy Policy Act is the newest and most comprehensive federal law addressing alternative transportation fuel use. Enacted on October 24, 1992, this law broadens the clean fuel program to include vehicle fleets operating in up to 150 cities. It requires the federal government to purchase at least 10,000 alternative fuel vehicles by 1995, with alternative vehicles comprising at least 75 percent of new purchases in 1999 and thereafter. Furthermore, the Energy Policy Act establishes grant and loan programs to pay the differential costs of alternative fuel vehicles compared with conventional vehicles. The law also creates a tax deduction of up to $2,000 available to the purchaser of each passenger vehicle operating on alternative fuels.[57]

State governments have acted on their own, and often more vigorously than the federal government, to promote the use of alternative transportation fuels. Since 1989, some 35 states have established alternative fuels programs, ranging from study groups to expansive fuel-switching mandates. Seventeen states now have laws mandating the use of alternative transportation fuels in specified fleet applications, usually state government vehicles. Other state government programs include vehicle demonstration projects, tax credits and exemptions, cash rebates, and incentive pricing for fuel.

The California Low-Emission Vehicle Program represents probably the most significant state government program promoting alternative transportation fuel use. Established in 1990, these regulations require the introduction of increasingly cleaner vehicles in four phases over a 15-year period. Full implementation of this program would lead to the introduction of zero-polluting electric vehicles to account for at least 10 percent of new automobile sales, rendering the conventional internal combustion engine obsolete, at least for a portion of the California market.

These federal and state government initiatives represent a serious commitment to begin the transition from oil-derived fuels as the virtual sole source of energy in the transportation sector. As of 1994, there were about 300,000 cars powered by liquefied petroleum gas in the United States, 50,000 natural gas vehicles, most of the world's nearly 15,000 methanol vehicles, about 3,000 battery-equipped electric vehicles, and about 500 vehicles running on ethanol.[58]

Compliance with the Clean Air Act Amendments alone is expected to create a market for more than one million alternative fuel vehicles by the turn of the century, compared with a current 190 million vehicles on the road. More optimistic estimates suggest that 20 million alternatively fueled vehicles, or nearly 10 percent of the current market, will be operating by 2010.[59]

An Increase in Natural Gas Vehicles

Of all the alternative transportation fuels now available to displace gasoline, natural gas has emerged as the most promising fuel option for the coming decades.[60] Natural gas is cheaper, cleaner, safer, and more abundant than oil. It is the only alternative transportation fuel that does not face some serious constraint – such as unavoidable high production costs or inescapable limitations on short-term supplies – that would undercut its viability as a major player in the automotive fuel market. The many advantages of natural gas include the following:

- **Environmental** Burning natural gas releases about 10 percent to 15 percent less carbon dioxide than burning gasoline. Hydrocarbon and nitrogen oxide emissions are also typically below pollution levels from gasoline, often as much as 40 percent to 80 percent less. Natural gas use reduces discharges of carbon monoxide and most toxic air chemicals dramatically, often by more than 90 percent. It also eliminates other pollution problems from oil production and use, such as oil spills and leaking underground gasoline storage tanks. The environmental benefits of natural gas exceed those achievable using ethanol or methanol – two competing alternative fuels.

- **Supply** There is more known natural gas in the world than there is oil, and, at current production rates, natural gas is being depleted at only half the rate of oil. Since industry has invested less effort in exploring for natural gas compared with oil, there is more reason to hope for discovery of new, larger natural gas deposits in the future. The current production capacity of natural gas in the United States could fuel 20 percent of the nation's vehicles – without diverting natural gas from other uses.[61] The ratio of known American reserves to current consumption is twice as high for natural gas as for oil.

- **Energy Security** The United States produces about 90 percent of its natural gas consumption domestically and imports nearly all of the remainder by pipeline from Canada.[62] Our other neighbor, Mexico, is an additional potential source. No other alternative fuel can be as easily and cheaply produced domestically as natural gas.
- **Economics** The price of producing natural gas is about half the price of gasoline. Adding distribution and refueling costs, natural gas "at the pump" sells at a nationwide average of 70 percent of the price of gasoline.[63] All other alternative fuels are more expensive to buy than gasoline.
- **Safety** Records from more than 25 billion miles driven worldwide in natural gas vehicles show that there has not been a single death from an accident involving a natural gas storage tank. By contrast, thousands of people die every year from burns received from gasoline fires due to gas tank failures during accidents. Liquid fuels, including gasoline, methanol, and ethanol, pool during a tank rupture and can cause fires. Moreover, the injury rate for natural gas pipeline workers is only a small fraction of the rate of injury to drivers of gasoline distribution tanks.

In the 1990s, the use of natural gas vehicle technology has experienced a dramatic era of growth in the United States and, to a lesser extent, worldwide. In the United States, annual growth rates in the number of natural gas vehicles have been in the double digits since 1990; about 50,000 natural gas vehicles were on US roads at the end of 1994. The number of natural gas refueling stations has doubled each year of this decade, reaching 1,000 in early 1995. Worldwide, there are now six countries, including the United States, with more than 20,000 natural gas vehicles operating. Three of these – Italy, Argentina, and Russia – each have more than 100,000 natural gas vehicles.

The number of organizations promoting natural gas vehicle use has also grown. The International Association for Natural Gas Vehicles was formed in 1986, while the United States Natural Gas Vehicle Coalition was founded in 1989. In late 1993, 63 companies from 17 nations in Europe joined to form the European Natural Gas Vehicle Association to promote natural gas vehicle use on that continent.

INFORM chronicled the growth of natural gas as a transportation fuel in its 1993 report *Paving the Way to Natural Gas Vehicles,* which further describes the benefits of natural gas as a transportation fuel. The report identifies the 25 main obstacles facing broader use of natural gas vehicles and profiles specific initiatives that address them. As of the major constraints for the shift away from oil-derived fuels is the lack of a refueling infrastructure, the report highlights the relative ease and benefits of using natural gas in urban fleets – particularly in those cities with the most polluted air.

The Need for Sustainable Transportation

The move to natural gas will undoubtedly reduce urban air pollution, extend the world's supply of crude oil, and provide additional time for the development and commercialization of transportation technology based on energy supplies that are non-polluting and sustainable. But its availability is limited to decades, not centuries, in the face of the world's huge and growing energy appetite. Whether postponed by natural gas or other currently available alternatives, that "day of reckoning" is unavoidable. Like oil, natural gas is a nonrenewable resource that will eventually disappear as a viable source of energy, and the environmental consequences of its use – for example, global warming – could necessitate limits on its use even before supplies dwindle. Although natural gas is much cleaner than oil-derived fuels, it is still a carbon-containing fossil fuel and its combustion releases greenhouse gases and some other air pollutants. Depending

Figure 6. A Sign for the Times?

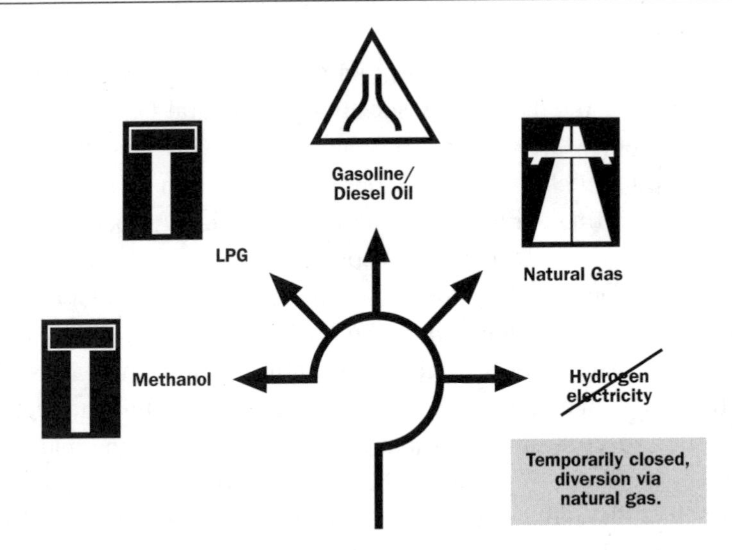

This drawing from the Australian Gas Association's *NGV Newsletter* shows dead ends for methanol and liquefied petroleum gas, a narrowing road for gasoline and diesel oil, and a bridge from natural gas to hydrogen.

on the pollutant of concern and the rate at which the automobile population increases in the future, growth in vehicle usage will ultimately erase the environmental advantages enjoyed by natural gas compared with oil use today.

Replacing oil with natural gas and other cleaner-burning alternative transportation fuels is certainly an important way to reduce pollution and alleviate the problems raised by oil dependence during the next few decades. However, the long-term challenge involves developing renewable energy resources to produce transportation fuels that are pollution-free from their creation to emission at the end of the tailpipe.

In order for an energy resource to provide a "centuries" solution to the world's growing transportation energy and environmental woes – rather than one for decades – it must be available in quantities so large as to be essentially limitless, and it must be tapped in a manner that produces so little pollution as to be essentially pollution-free. The term "sustainable" is frequently used to reflect these two attributes of an energy resource. An expanding energy system that meets these two criteria of energy sustainability also exemplifies sustainable development which, as defined by the World Commission on Environment and Development in 1987, is development that "meets the needs of the present without compromising the ability of future generations to meet their own needs."[64]

In practical terms, only energy resources that renew themselves at least as quickly as they are consumed meet the criterion of being virtually limitless. Otherwise, it is only a matter of time before usage exhausts the resource. Similarly, resources meet the pollution-free criterion if their use emits so little pollution that the natural cleansing processes of chemical decomposition eliminate the pollutants before they accumulate in the environment. Otherwise, environmental damage, even if minimal at current levels of resource use, may eventually extract a toll.

The principles of energy sustainability – unlimited supplies and absence of pollution – must be maintained throughout the entire cycle of energy production, distribution, and use, generally termed the "fuel cycle." The promise of sustainable energy for transportation applications relies on two interrelated factors: the development of solar energy as a transportation energy resource and the development of clean energy "carriers" capable of distributing sustainable energy for use in vehicles. Electricity is the most common zero-polluting energy carrier in use today – its use produces no cumulative environmental impact. However, hydrogen may well be another important energy carrier of the future – different from but in many ways a complement to electricity.

Solar Energy for a Sustainable Future

Solar energy, in its various forms, is in a position to become the predominant energy resource of the sustainable energy era. For the last 4.5 billion years, the sun, now entering middle age, has lit the skies and warmed the earth without fail. The heat of the sun is generated by the fusion of over 500 million tons of hydrogen atoms every second. [65] In contrast, a mere 40 million tons of hydrogen are manufactured on Earth every year.[66] Although only 2.2 billionths of the total heat released by the sun reaches the Earth, the energy it delivers is nevertheless astounding. Every 30 minutes, more solar energy strikes the earth than is released from the burning of fossil fuels in a year.[67] Stated another way, the solar energy striking the earth is close to 20,000 times greater than the energy being released worldwide from burning fossil fuels. Tapping the sunshine falling on just 5 percent of the world's desert regions would supply enough energy to meet total world energy demand. Only 1 percent of the desert areas in the United States would be needed to supply enough hydrogen to power all the vehicles projected to be on American roadways in 2010.[68]

The Forms of Solar Energy

Solar energy already contributes about 15 percent of the world's energy supply, two-thirds in the form of biomass fuels and one-third in the form of hydropower – although none of this is used for transportation. When developed to even a small fraction of its potential, however, solar energy, because it is replenished daily, can meet future world demand for thousands of years to come. There are some environmental consequences of using the various forms of solar energy, but compared with fossil fuel use, solar energy production is extremely clean and, in most applications, causes virtually no air or water pollution.

Several renewable resources, all forms of solar energy, can be used to generate the electricity needed for large-scale hydrogen production: solar thermal, photovoltaic cells, wind, hydropower, and biomass. This solar-generated hydrogen could then power a pollution-free transportation system.

Solar thermal power

Among the many forms of solar energy, sunlight is the most obvious. When sunlight strikes a dark surface, heat is produced. Tapping this heat, called solar thermal energy, to heat water or to warm the interior of a building are commonplace applications of solar energy today. In an industrial modification of solar thermal technology, the sun's rays are focused on a single point to generate great heat. Like the heat produced from the burning of fossil fuels, concentrated solar rays can heat water in a conventional power-generating boiler. The resulting steam can then be used directly in industrial processes that require steam, or it can be used to generate electricity in a power turbine. There are close to 500 megawatts of thermal power generation capacity in the world today.

Photovoltaic cells

Sunlight can also be converted directly to electricity by solar photovoltaic cells. In this technology, the energy in sunshine is transferred to electrons via special photosensitive chemicals in the cells, triggering a flow of electricity. The total world electrical generating capacity from solar photovoltaic technology is only about 1,000 megawatts, equal to about one average-sized fossil fuel power plant. Tiny arrays of photovoltaic cells capable of powering watches and calculators represent the largest market. Larger arrays of photovoltaic cells power electrical equipment in remote areas not presently connected with electrical transmission lines. Globally, about 60 megawatts of electrical generating capacity from photovoltaic cells is marketed annually.

Photovoltaic cells have been the fastest growing market among the various types of solar energy technologies. Between 1976 and 1983, the worldwide photovoltaic market grew at an annual rate of 70 percent; since 1983 the rate has dropped to 16 percent per year, largely due to the falling price of oil and the cutback in government subsidies for renewable energy development.[69] Even this modest rate, however, is still more than five times greater than the 3 percent annual total growth rate for all forms of electrical generation.

Wind power

When sunlight strikes the earth and molecules of air in the atmosphere, a gradual warming results. This warming is not uniform; the ground heats faster than the air, for example. The uneven heating of the atmosphere creates wind currents, one of several indirect forms of solar energy. Windmills convert wind energy to mechanical or electrical energy. With current technology, it is cheaper to tap wind power to generate electricity than it is to use sunlight in solar thermal or photovoltaic technologies. There are now more than 17,000 electrical-generating wind machines operating in the United States, mostly in California, generating more than 1,500 megawatts of power. This is enough to satisfy the electricity demand of a city of 1 million people. Furthermore, there are hundreds of thousands of windmills in rural locations worldwide that use wind to pump groundwater to the surface. The United States has abundant wind resources, especially in the Great Plains states. Exploitable wind resources have been estimated recently to be four times greater than the total electrical demand rate; the wind potential of just two states, North and South Dakota, is about equal to the nation's total electrical generation.[70]

Hydropower

The evaporation of water from the earth's surface is another result of the warming effect of sunlight striking the earth. The resulting clouds cool as they rise and eventually produce rain, especially at higher altitudes. The rain collects in lakes and streams and begins the downhill journey to lower altitudes. Hydropower technology can tap the energy contained in this falling water. Because solar heat produces the energy needed to evaporate the water and to move it to the heights from which it naturally flows with the force of gravity, hydropower, too, is a form of solar energy. Hydropower represents the largest use of renewable energy technology today. About 3 percent of the total US energy supply and 6 percent of the world energy supply is obtained from hydropower facilities.

Biomass

All plant life, termed biomass, can be considered another indirect form of solar energy. Plants use the energy from the sun to convert carbon dioxide from the atmosphere and water and nutrients from the soil into wood and leaves. The burning of biomass releases this captured solar energy, returning the carbon dioxide and water back to the environment as part of a natural, closed cycle. Globally, biomass resources provide about 10 percent of all energy; in nonindustrialized countries, the figure is 38 percent.[71] Industries use about two-thirds of all biomass energy, mainly the timber products industry, which burns waste wood as a fuel. About 4 million homes in the United States rely on wood-burning as the primary source of heat. A third application of biomass is the conversion of corn into ethanol, which is blended with gasoline as an automotive fuel. About 10 percent of all gasoline sold in the United States is a blend containing 10 percent ethanol derived from corn.[72]

Drawbacks of Solar Energy

The diverse forms of solar energy suffer from two drawbacks when considered as substitutes for gasoline in transportation. First, they are very diffuse, so large land areas must be covered with energy collection systems in order to gather sizable amounts of energy. For example, on a sunny day the amount of solar energy falling on a square foot of land in the southwestern desert has less energy content than a lump of coal that easily fits in the palm of a hand. In addition, current solar energy technologies are able to capture only a small portion of the sun's energy, typically about 10 percent in the case of photovoltaic cells and far less than that for biomass. Technologies on the horizon should boost efficiencies of some solar technologies to more than 20 percent. At that efficiency, it would be necessary to cover about 3 percent of the land in the United States with solar energy collectors to supply all of the country's energy needs. Although small on a percentage basis, 3 percent nevertheless is roughly equal to the size of the state of Wyoming.[73]

Developing a solar energy system on this scale is obviously a major undertaking. A key objective of solar energy research today is to improve the efficiency of solar collection. This would reduce the cost of solar technologies and reduce the land requirements needed to capture this diffuse energy resource.

The second drawback is that solar resources are unequally distributed throughout the world in a physical and temporal sense. Prime solar energy fields are often found in sparsely populated, remote regions. Areas with strong sunlight, termed high solar insolation, include deserts; high wind resources are found most commonly in mountain passes and high plains; biomass resources are abundant in wooded areas, and hydropower resources are most abundant in tropical rainforests. Moreover, even when found in high concentration, solar energy resources are intermittent in their availability: direct sunlight is, obviously, available only during daylight hours. Similarly, wind resources can be tapped only on windy days, and hydropower's reliability is subject to droughts.

Unlike oil, solar resources cannot be packaged in their naturally occurring form and shipped to where the demand for energy exists: they must be tapped where and when they are available. In order for solar energy resources to become practical as sustainable transportation energy resources, their energy must first be converted into some other form that is easy to store, transport, and use.

Because of these limitations, solar energy is not now a significant provider of transportation fuels. Sailboats, river barges floating downstream, and futuristic cars powered by roof-mounted photovoltaic cells are exceptions. Sailboats, however, languish in calm weather, barges without tugboats are limited to downstream destinations, and solar-powered cars cannot run for long at night, in darkened tunnels, or on cloudy days.

Energy Carriers in a Sustainable Transportation System

The answer to the problem of using diffuse, remote, and intermittent solar energy resources in transportation applications lies in the development of energy "carriers," concentrated forms of energy that can be moved about easily. Carriers are transformed energy resources; large supplies do not occur naturally on Earth but rather are created from other energy resources. The challenge in transportation is to develop the carrier or carriers that can best deliver solar-derived fuel to vehicles.

Electricity: The World's Major Energy Carrier

The energy carrier that most people experience daily is electricity. With the exception of brief flashes of lightning and sparks of static electricity, electricity is not found naturally in high concentrations. But electricity can be readily produced from other energy resources at power-generating plants, where the energy that fuels the power plant is converted into electrical energy. Once generated, electricity can be transported through transmission lines to the point where it is used to power appliances, lights, computers, cars, etc.

The transmission and use of electricity is nearly pollution-free. It has no smell, emits little heat, makes no noise, and weighs nothing. Distribution and use of electricity, therefore, is environmentally sustainable. Electricity performs many functions, such as producing illumination from a light bulb, that cannot be performed as effectively by the energy resources from which the electricity was originally produced. An electric bulb, for example, produces more illumination than an oil-burning lamp consuming the same amount of energy. Electricity can do some things, such as storing information in a computer or receiving radio and television signals, that other energy resources cannot do at all.

It is no wonder that nearly one-third of global energy production is used to generate electricity.[74] Since 1892, when the world's first power-generating station began providing electricity in New York City, the spectacular growth in the use of electricity and the impact of this new energy resource on the way we live has paralleled the growth in gasoline-burning automobiles during the same period.[75] Nearly all of the world's uranium production is used to generate electricity. More than half the world's coal production and a quarter of its natural gas production is likewise used to generate electricity that is "carried" to energy markets by transmission lines. Solar photovoltaic cells produce electricity as their only product, and electricity is a common product from tapping solar thermal, wind, hydropower, and biomass energy resources.

Although vehicles powered by electric batteries enjoyed a brief moment in automotive history around 1900, the proliferation of electricity and automobiles have until recently been distinct, non-overlapping occurrences. Despite the almost magical allure of electricity as an energy carrier and its central role in improving the quality of 20th century life, the application of electricity in automotive transportation presents several challenges.

First, while the distribution and use of electricity are sustainable, its production is not necessarily so. Today, nearly two-thirds of the electricity generated worldwide uses nonrenewable fossil fuels as the primary energy resource; in the United States, the figure is 58 percent. Nuclear power, which also uses nonrenewable resources, generates another 17 percent of worldwide electricity and 21 percent of American consumption.[76] With its reliance on fossil and nuclear fuels, electrical generation is currently a major source of air pollution. In the United States, for example, electrical generation produces 72 percent of the nation's total emissions of sulfur dioxide, 33 percent of the nitrogen oxide emissions, and 10 percent of the particulate emissions.[77] Added to these impacts are the environmental effects of mining coal and uranium, water pollution, and solid waste impacts, including the disposal of radioactive wastes from nuclear generation. Electrical generation from nonrenewable resources is also an energy-inefficient technology: almost two-thirds of the energy in primary energy resources is lost during the generation of electricity.[78]

The emergence of a sustainable electricity future relies on the transition in electrical generation from nonrenewable, polluting fuels to solar energy resources. Even when electrical generation is produced from renewable resources, however, its transmission still causes some problems: electrical shocks involving high voltages and currents, which are always dangerous and sometimes fatal; electromagnetic fields near power lines, which some evidence suggests can cause adverse health effects, including leukemia and other types of cancer; and the visible intrusion presented by large overhead power lines extending for thousands of miles, which is a significant land-use impact that is difficult to mitigate. Further, since stringing transmission lines across oceans is impractical, the energy carrying capabilities of electricity are limited to the land masses where it is generated – unlike oil, which is distributed globally by oceangoing tanker ships.

The biggest problem with electricity, however, is the difficulty of storing it. Virtually the instant that electricity is generated, it is consumed in millions of end-use applications, often located hundreds of miles away from the relatively small number of power plants. As demand for electricity fluctuates at the point of use, the output of power plants is simultaneously adjusted so that the supply of electricity almost exactly matches the demand. To achieve this balance, an assortment of power plant technologies have been developed and commercialized that differ in their ability to respond to changes in electrical demand:

- Baseload power plants are most efficient and inexpensive when operated at full capacity around the clock. The number and size of baseload power plants, therefore, is matched to meet the minimum continuous, or base, demand for electricity.
- Intermittent power plants produce electricity that is slightly more expensive than the output of baseload facilities. Their advantage is that they can be rapidly turned on and off.
- Peaking facilities are built to provide electricity only during the few hours each year when electricity demand skyrockets – for example, during heat waves. Peaking plants are typically the cheapest to build, but they are also the most inefficient and expensive to operate.

The process by which intermittent and peaking power plants are started up and shut down in the face of changes in demand for electricity is called "dispatch." Thousands of changes are made in the mix of power plants used in the United States during the course of a day in response to dispatch decisions. The complex system of electrical dispatch is needed because electricity is very difficult to store. Like fire, it is a phenomenon, not a substance. It cannot be held in a tank until needed, like gasoline, nor can it be stockpiled like coal. There are technologies that can store electricity in an alternative form temporarily and then reconvert it to electricity when needed, but the dispatch system is cheaper and more efficient.

Electrical storage is needed, however, when it is impractical for an electrical device to be continuously connected to a transmission line. Flashlights are one obvious example where electrical storage is obligatory; electric automobiles are another. Automobiles must carry enough energy on board to have a reasonable driving range between refueling stops. In order to compete with conventional automobiles, electric vehicles must store enough electricity to propel a vehicle for several hundred miles without recharging.

Batteries are currently the most widely used electrical storage technology, but they are heavy, bulky, and sometimes filled with toxic chemicals. Batteries capable of storing an amount of electricity equal to the energy in a couple of gallons of gasoline can weigh more than one ton. Replenishing depleted batteries with new stores of electricity is time-consuming, and the number of times recharging can occur is limited by chemical decomposition. Once recharged, the flow of electricity from the batteries to the electric motor powering a vehicle is often too slow or too weak to provide enough power to match the performance of a gasoline-burning engine.

An analysis of electric cars publicized in a 1903 issue of the automotive magazine, *Overland Monthly,* observed: "The many drawbacks of such machines today are the smaller radius of action, slow speed, weight of batteries, and cost and time taken to recharge."[79] Problems with energy storage were a key reason for the demise of battery-equipped electric vehicles at the turn of the century in favor of gasoline-powered cars, and despite significant progress, especially in recent years, battery storage remains today as a key obstacle to the reemergence of electric vehicles.

During the past decade, hundreds of millions of dollars have been invested in developing battery technology capable of powering electric vehicles. Research is also underway to develop alternative storage systems such as flywheels, which hold electricity in gyroscopic devices that spin rapidly in a magnetic field. The results have been both impressive and widely publicized. Prototype electric vehicles are now entering the marketplace capable of performing well in applications that do not require extended driving between refueling. Despite this progress, significant breakthroughs in battery technology are required before battery-powered electric vehicles can match the cost and performance of conventional gasoline-burning automobiles.

Electric vehicles powered by electricity stored in batteries clearly can play a major role in reducing automotive emissions and providing an alternative to gasoline-powered vehicles – even without additional technological breakthroughs. The role could be limited, however, to using batteries in low-mileage vehicles driven largely in heavily polluted areas. They may also play a role as a transitional technology used briefly until better options are developed.

Hydrogen: A New Energy Carrier for Transportation

The limitations on battery storage of electricity have prompted the search for energy carriers other than electricity and for energy storage technologies other than batteries that may prove to be viable components of a sustainable transportation system. Hydrogen is such a sustainable energy carrier. It has all the traits a sustainable energy carrier needs: It is storable, movable, usable in a wide variety of applications, and nonpolluting. Moreover, it can be produced using solar energy resources. Therefore, like electricity produced from solar energy, solar-derived hydrogen can form the basis of a sustainable transportation energy future. And because electricity can be used to produce hydrogen from water, and hydrogen can be used to generate electricity in a fuel cell, the use of these two energy carriers can be linked to allow each to play an optimal role for sustainable transportation.

Hydrogen is the most abundant element in the universe, forming about 80 percent of all matter.[80] Each of the billions of stars, for example, is a flaming orb of hydrogen. Yet like electricity, hydrogen does not exist in pure form on this planet. Hydrogen atoms are too reactive to occur as single, unattached units; instead, they easily combine with many chemical elements, including other hydrogen atoms, to form chemical compounds. Most known molecules, including all the fossil fuels, contain some hydrogen atoms. In fact, much of the energy released by the burning of fossil fuels results from breaking chemical bonds involving hydrogen atoms.

Molecular hydrogen forms when two atoms of hydrogen combine. Molecular hydrogen is a very potent energy resource. Burning hydrogen molecules releases 3.2 and 2.8 times more energy per pound than gasoline or natural gas, respectively. Molecular hydrogen, however, is also one-seventh the weight of air and, because of this, virtually all naturally occurring molecular hydrogen has long since escaped Earth's atmosphere and drifted into interstellar space. Less than 0.2 percent of the atmosphere is molecular hydrogen gas, most of which is found at higher altitudes.

In order for molecular hydrogen to be used as an energy carrier, it must, like electricity, first be created on Earth. This can be accomplished through chemical reactions that extract hydrogen from hydrogen-containing molecules. The most common method of producing hydrogen today is through a reaction between natural gas and water, called "steam reforming." Such reactions invariably require energy and often produce some pollution, although less than the pollution produced from burning fossil fuels to generate electricity.

There is a sizable hydrogen production industry in the world today, although it is dwarfed by the markets for energy. About 40 million tons of hydrogen are now produced annually worldwide, or about 15 trillion cubic feet of hydrogen gas. This hydrogen contains about 5 quadrillion British thermal units (Btu) of energy, or slight more than 1 percent of the world's energy demand of 385 quadrillion Btu.[81] The United States produces about 9 million tons of hydrogen – about a quarter of worldwide production – but this supplies only 1.2 quadrillion Btu. In comparison, the United States produces about 20 quadrillion Btu of natural gas and consumes a total of about 82 quadrillion Btu.[82]

Only a tiny fraction of the world's hydrogen production is used as an energy resource. Chemical manufacturing and oil refining, not energy production, currently make up the principal markets for hydrogen. In the United States, roughly 60 percent of all hydrogen produced is used in oil refining; 30 percent is used in ammonia production and ultimately incorporated into fertilizers; and about 10 percent is used for other purposes, including metal production, food processing, and rocket propulsion. Most hydrogen – more than 98 percent – is consumed by the companies that produce it; this is termed "captive hydrogen production." "Merchant hydrogen," meaning hydrogen sold on the open market, accounts for less than 2 percent of American production. The space program is the largest consumer of merchant hydrogen, a key fuel used to propel spacecraft into orbit. Except for the space program, which purchases some 20 tons of hydrogen daily, virtually no hydrogen today is used in transportation or other energy applications.[83] Worldwide ammonia production accounts for nearly two-thirds of all hydrogen use, and oil refining consumes another 10 percent.[84]

A Sustainable Hydrogen Transportation System

Even if current methods of producing hydrogen were greatly expanded, this would not support a sustainable hydrogen-based transportation system because today's hydrogen production technologies rely on fossil fuels. For hydrogen to be used as an energy carrier in a sustainable transportation system, technologies must be perfected that can produce it from renewable resources, and automotive systems must be developed that can use it as a transportation fuel. The most technologically advanced and extensively studied method of producing hydrogen from a renewable resource is to split a molecule of water into its constituent elements, two atoms of hydrogen and one atom of oxygen.

The British scientist Henry Cavendish first isolated hydrogen in this fashion in the 1760s and identified hydrogen's principal characteristics. The name "hydrogen" in Latin literally means "produced from water." In the 1780s, Cavendish showed experimentally that igniting a mixture of hydrogen and air with a spark produced an explosion and a small quantity of water.[85]

Cavendish discovered the chemical principles for what has evolved into the concept of a sustainable "hydrogen energy economy." First coined by energy analysts in the early 1970s, the sustainable hydrogen energy economy begins with the production of hydrogen from the splitting of water using solar energy. Once produced, hydrogen can be stored in compressed form and transported efficiently through pipelines. Alternatively, it can be liquefied and transported in ships across oceans. Then it can be pumped in compressed or liquefied form into fuel storage tanks.

When energy is required, hydrogen can be burned like any other fuel to generate heat. As energy is released, the hydrogen chemically recombines with oxygen to form water, thereby completing the hydrogen fuel cycle with little or no air pollution. Hydrogen can also be used in fuel-cell engines – the same technology that powers spacecraft. Fuel cells produce electricity from the electrochemical recombination of hydrogen and oxygen, without combustion. Vehicles powered by hydrogen fuel cells, therefore, are a type of electric vehicle. While a battery-powered electric vehicle stores electricity generated elsewhere in on-board batteries, a fuel-cell-powered electric vehicle actually generates electricity on board, from hydrogen.[86] The energy contained in the hydrogen in 1 gallon of water can propel a fuel-cell-powered hydrogen vehicle as far as 1 gallon of gasoline propels a conventional car.[87]

This vision of sustainable transportation system based on hydrogen thus consists of three steps:

1. Tapping solar energy to produce hydrogen from water.
2. Using hydrogen as an energy carrier to transport and store energy for use in vehicles.
3. Releasing the energy in hydrogen on board a vehicle by recombining hydrogen and oxygen to regenerate water, either by burning the hydrogen or by using it to generate electricity in a fuel cell.

Electrolysis, using electricity generated from solar energy, is one common way to produce hydrogen. There are other sustainable energy systems that do not include electrolysis. For example, hydrogen can be obtained from plants, called biomass, which themselves originally captured water and solar energy as they grew. Some bacteria and algae also produce hydrogen naturally, again, ultimately using solar energy and water to accomplish this.

Although hydrogen-powered vehicle technology is less developed than electric battery-powered vehicle technology, hydrogen has been successfully demonstrated as a viable transportation fuel over a period of more than a century. Thousands of hydrogen-burning engines were built in Germany before and during World War II. Scores of conventional vehicles have been converted to run on hydrogen for testing purposes during the past 20 years. Recently, great strides have been achieved in developing fuel cells suitable for use in vehicles.

The Hydrogen – Electric Vehicle Connection

The distribution and use stages for both hydrogen and electricity are sustainable. The pipeline distribution of hydrogen and the transmission of electricity through power lines occur without creating cumulative environmental impacts. The use of electricity is similarly pollution-free. And when hydrogen is used in a fuel cell to generate electricity, its use, too, is pollution-free: the sole product of hydrogen fuel cells is water. When hydrogen's energy is tapped through burning, the amount of air pollution that is produced is so small as to be of little concern, even if huge amounts of hydrogen are consumed.

There are also analogies between hydrogen fuel cells and electric batteries. Electricity can be stored in batteries in the form of chemical energy and can then be reconverted to electricity on board the vehicle. Electricity can also be used as a source of energy to produce hydrogen by splitting the water molecule. This hydrogen can be stored in gaseous or liquid form and then reconverted to electricity in a fuel cell. Hydrogen fuel cells and batteries, therefore, offer two different ways of storing electricity and providing it to power electric vehicles. Although the mechanics of the systems are different, vehicles can be equipped with both fuel cells and batteries, resulting in a type of vehicle called a "hybrid." Most electric components now used in electric vehicles powered by batteries, including the motor, controllers, and braking systems, are equally usable in vehicles powered by fuel cells.

Links between Hydrogen and Natural Gas Vehicles

Hydrogen offers a sustainable transportation option for the future because it can be produced from solar energy and used in electric vehicles. Hydrogen also has links to conventional automotive technology and to a leading alternative transportation fuel gaining in popularity today: natural gas. Like natural gas, hydrogen can be burned in conventional internal combustion engines. In fact, most of the engine modifications needed to switch from burning gasoline to natural gas are required to burn hydrogen as well, and cars are now being tested that burn a mixture of hydrogen and natural gas called "Hythane®."[88] Onboard storage systems for natural gas and hydrogen are analogous, as is refueling technology. Thus, technological advancements in natural gas vehicle and refueling technologies are likely to be directly applicable to hydrogen vehicles.

As mentioned earlier, current hydrogen production methods are linked to natural gas; hydrogen is manufactured today almost exclusively from natural gas. Moreover, many natural-gas-producing areas are located in regions, such as the Southwest, which are rich in solar energy. As natural gas fields are depleted in the future, these sites could serve as locales for hydrogen production from solar energy.

Fuel distribution systems for natural gas and hydrogen are also similar. Pipelines offer the cheapest and easiest mode of long-distance land transportation for both fuels. In fact, a mixture that is 20 percent hydrogen and 80 percent natural gas can be transported through natural gas pipelines without need for any equipment modifications. Ultimately, existing natural gas pipelines could then be modified to carry the hydrogen to energy markets or replaced with new hydrogen pipelines along the same pipeline right-of-ways. Liquefaction of both natural gas and hydrogen, and shipment in specially insulated tankers offers a viable means of moving either fuel across oceans.

Hydrogen faces many of the same obstacles that impede the use of natural gas as a transportation fuel. It does not suffer, however, from the two major drawbacks of natural gas: limited supply as a fossil fuel and the release of some air pollution upon combustion. A synergy between use of natural gas and hydrogen may exist if a transition to natural gas vehicles helps pave the way for hydrogen by providing the infrastructure and technology needed to use gaseous fuels in transportation. While natural gas vehicles are part of the solution to our transportation crisis for the coming decades, hydrogen provides an energy resource that essentially can last forever. The role of natural gas in helping to establish the ultimate availability of hydrogen as part of a sustainable transportation system – based on hydrogen production from solar energy and use in fuel-cell-powered electric vehicles – provides a critical policy rationale to justify the tremendous investment needed to create a large natural gas vehicle market. Thus, as discussed further in Chapter 5, natural gas can serve as bridge to a sustainable transportation energy system – even though, on its own, it is not a long-term solution.

Transportation Energy Pathways

The world is in the middle of a second great transportation revolution. Like the move from horses to automobiles a century ago, the current revolution will result partly because the present way of doing things cannot continue and partly because a better way of doing things is coming along. inform's question is: What role might hydrogen play in a sustainable transportation future, and how will the transition to that future unfold?

Figure 7. Transportation Energy Pathways

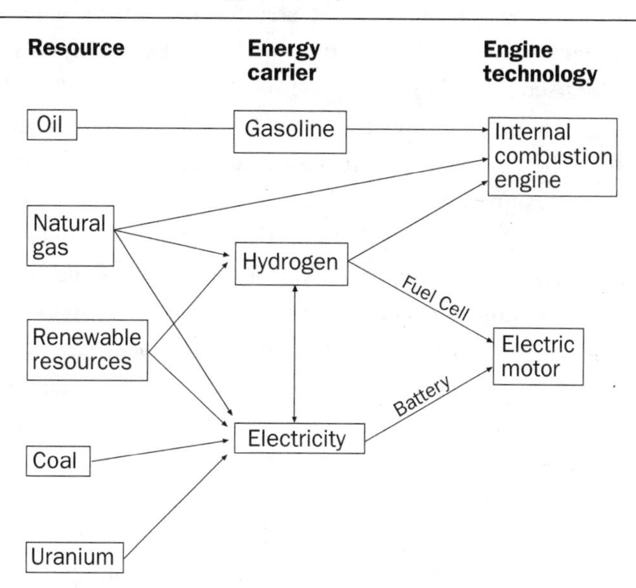

Note: Additional candidates for use in transportation energy systems include other nuclear fuels for fusion reactors and other energy carriers, such as propane, methanol, and ethanol.

Source: INFORM

Figure 7 summarizes the leading actors in today's transportation revolution. The left side lists the primary energy resources; the nonpolluting energy carriers – hydrogen and electricity – appear in the middle; and the two automotive propulsion options – the internal combustion engine and the electric motor – are along the right. The figure presents the scenarios for the production and application of energy for transportation that are in use today, as well as those that are most actively under development and competing for a place in transportation systems of the future. The figure does not show every conceivable pathway, but focuses rather on those that appear most viable.

The status quo is represented at the top. Oil supplies gasoline that is burned in automobiles powered by internal combustion engines. Other fuels are competing with gasoline, however. Most prominent among them is natural gas. Hydrogen can also be burned in internal combustion engines, and one energy pathway represents the steam reforming of natural gas to produce hydrogen that is burned in internal combustion engines. Hydrogen can also be used in fuel cells to generate electricity capable of powering electric vehicles. Therefore, a second line connects hydrogen to the electric motor. The primary resource for hydrogen may also include renewable energy resources as well as natural gas.

Electricity, the other major energy carrier in transportation systems of the future, can also be generated from various sources, including coal and uranium as well as natural gas and renewable energy resources. Oil is too limited and costly for widespread use in generating electricity. Electricity can be used only to power electric motors; it cannot be used to power internal combustion engines. The only link between electricity and internal combustion engines is through use of electricity to produce hydrogen. This link is a two-way street, because hydrogen can also be used to generate electricity. There is a strong synergy between the electric-vehicle technology used in a battery-powered vehicle and the technology applicable to fuel-cell-powered vehicles.

A key point demonstrated in Figure 7 is the versatility of both natural gas and hydrogen in possible transportation energy futures. Natural gas can be burned directly in the engine of today's cars; it is also a major fuel in the generation of electricity and is the source of most hydrogen production. Hydrogen shows a similar flexibility. It is capable of powering both internal combustion engines and electric motors; it can generate electricity and be produced from electricity; and it can be produced directly from other energy sources as well. Hydrogen's central position in many plausible transportation energy pathways provides much of the basis for INFORM's focus on hydrogen as a fuel of the future.

Notes

1. J. Ausubel, "Hydrogen and the Green Wave," *Proceedings of the First Annual Meeting of the National Hydrogen Association (*Washington, DC, March 8-9, 1990), pp. 20-13 and 20-23.
2. Automobile registrations from Daniel Yergin, *The Prize (*Touchstone Press, New York, 1991), p. 80.
3. The main reference for most of this capsule summary of the early days of the oil industry was: Yergin, *The Prize, op. cit.*
4. Exhibits at the Owls Head Transportation Museum in Rockland, ME, visited by INFORM, June 14, 1994; and C. McShane, "The Combustion When Cars First Hit the Streets," *Newsday,* May 4, 1994.
5. *Yergin, op. cit.,* p. 80; and James MacKenzie, *The Keys to the Car: Electric and Hydrogen Vehicles for the 21st Century (*World Resources Institute, Washington, DC, 1994), p. 33.
6. Exhibits at the Owls Head Transportation Museum, *op. cit.*
7. McShane, *op. cit.*
8. D. Sperling, *Future Drive: Electric Vehicles and Sustainable Transportation* (Island Press, Washington, DC, 1995), p. 36.
9. Deborah Gordon, *Steering a New Course: Transportation, Energy, and the Environment,* (Union of Concerned Scientists, Cambridge, MA, 1991), pp. 5-10; and Steve Nadis and James MacKenzie, *Car Trouble* (World Resources Institute, Washington, DC, 1993), pp. 2-5.
10. American Automobile Manufacturers Association, *Facts & Figures 1993* (Detroit, MI, 1994), p. 21.
11. *Ibid.,* pp. 3 and 64.
12. *Ibid.,* pp. 29 and 51.
13. L. Schipper *et al.,* "Energy Use in Passenger Transport in OECD Countries: Changes Between 1970 and 1987," *Transportation, The International Journal,* April 1992, cited in Congress of the United States, Office of Technology Assessment, *Saving Energy in U.S. Transportation* (Washington, DC, July 1994), p. 1.
14. American Automobile Manufacturers Association, *World Motor Vehicle Data: 1993 Edition (*Detroit, MI, 1994), pp. 3 and 23.

15. American Automobile Manufacturers Association, *Facts & Figures 1993*, *op. cit.*, p. 39.

16. Stacy Davis, *Transportation Energy Data Book: Edition 14* (Oak Ridge National Laboratory, Oak Ridge, TN, 1994), pp. 3-11 and 4-2; and American Automobile Manufacturers Association, *Facts & Figures 1993, op. cit.*, p. 64.

17. American Automobile Manufacturers Association, *World Motor Vehicle Data: 1993 Edition, op. cit.*, pp. 25-27.

18. INFORM calculations based on data from Yergin, *The Prize, op. cit.*, pp. 793 and 830; and United States Energy Information Administration, *International Energy Annual 1992* (Washington, DC, 1994), pp. 6 and 105.

19. R. Riley, *Alternative Cars in the 21st Century* (Society of Automotive Engineers, Warrendale, PA, 1994), p. 10.

20. United States Energy Information Administration, *Annual Energy Review 1993* (Washington, DC, 1994), p. 141.

21. *Ibid.*, pp. 141 and 163.

22. "Taking an Alternative Route," prepared by Argonne National Laboratory for the United States Department of Energy (Washington, DC, 1994), p. 2.

23. United States Energy Information Administration, *Annual Energy Outlook 1992* (Washington, DC, 1992), pp. 20 and 27.

24. Domestic Fuels Alliance, press release, (Washington, DC, March 15, 1995).

25. C. Moore and A. Miller, *Green Gold* (Beacon Press, Boston, MA, 1994), pp. 60-62.

26. United States Energy Information Administration, *Annual Energy Review 1993, op. cit.*, p. 163.

27. C. Masters *et al.*, "World Oil and Gas Resources – Future Production Realities," *Annual Review of Energy,* Vol. 15, 1990, pp. 23-51, cited in R. Williams and G. Terzian, "A Benefit/Cost Analysis of Accelerated Development of Photovoltaic Technology" (Princeton University Center for Energy and Environmental Studies, Princeton, NJ, October 1993), p. 1.

28. Stephan McCrea, "Preparing for the Post-Petroleum Era: What Policy Analysts and Administrators Need to Know About Alternative Fuel Vehicles," *Environmental and Urban Issues,* July 1994, p. 17.

29. Christopher Flavin and Nicholas Lenssen, "Sustainable Energy for Tomorrow's World" (Worldwatch Institute, Washington, DC, 1994), p. 3.

30. American Automobile Manufacturers Association, *Facts & Figures 1993, op. cit.*, p. 83.

31. *Ibid.*, p. 88.

32. United States Environmental Protection Agency, *National Air Quality and Emissions Trends Report 1993* (Washington, DC, October 1994).

33. James S. Cannon, *The Health Costs of Air Pollution* (American Lung Association, Washington, DC, 1990).

34. "Bad Air May Cut Life Span," *Albuquerque Journal,* Albuquerque, NM, December 9, 1993.

35. International Energy Agency, *Greenhouse Gas Emissions: The Energy Dimension* (Paris, France, 1991), p. 49.

36. United States Energy Information Administration, *Emissions of Greenhouse Gases in the United States 1987-1992* (Washington, DC, November 1994), pp. 9 and 12.

37. International Energy Agency, *Greenhouse Gas Emissions: The Energy Dimension, op. cit.,* pp. 46-49. A thorough analysis of climate change associated with transportation appears in another IEA book, *Cars and Climate Change,* published in 1993.

38. Dr. Joel Levine, "Global Climate Change," a chapter from *Global Climate Change and Freshwater Ecosystems,* Penelope Firth and Stuart Fisher (ed.), (Springer-Verlag, New York, 1993); and background materials sent to INFORM by Dr. Levine, April 1994.

39. Moore and Miller, *Green Gold, op. cit.,* p. 71.

40. William Clinton and Albert Gore, Jr., *The Climate Change Action Plan* (The White House, Washington, DC, October 1993).

41. United States Energy Information Administration, *Annual Energy Review 1993, op. cit.,* p. 87.

42. United States Energy Information Administration, *Annual Energy Outlook 1992, op. cit.,* p. 27.

43. United States Energy Information Administration, *Annual Energy Review 1993, op. cit.,* p. 141.

44. American Petroleum Institute, press release, January 18, 1995, cited by Domestic Fuels Alliance, press release (Washington, DC, March 15, 1995).

45. United States Energy Information Administration, *Annual Energy Outlook 1992, op. cit.,* p. 27.

46. Domestic Fuels Alliance, "Petroleum Subsidies" (Washington, DC, March 13, 1995).

47. Mark DeLuchi, "Hydrogen Vehicles: An Evaluation of Fuel Storage, Performance, Safety, Environmental Impacts, and Cost," *International Journal of Hydrogen Energy,* Vol. 14, No. 2, February 1989; American Automobile Manufacturers Association, *Facts & Figures 1993, op. cit.,* p. 83; and "Current Emission Control Costs Calculated," *Clean Fuels Report,* April 1993.

48. James S. Cannon, "Reformulated Gasoline: Cleaner Air on the Road to Nowhere," INFORM, New York, 1994.

49. C. Flavin and N. Lenssen, *Power Surge: Guide to the Coming Energy Revolution* (W.W. Norton & Company, New York, NY, 1994), p. 69.

50. A thorough discussion of environmental externalities from transportation fuel use appears in Chapter 4 of *Saving Energy in U.S. Transportation,* Congress of the United States, Office of Technology Assessment (Washington, DC, July 1994) and in Chapter 7 of this INFORM report.

51. United States Energy Information Administration, *International Energy Annual 1992, op. cit.,* p. 105.

52. Figures for Hiroshima and Nagasaki from *Encyclopedia Americana International 1994* (Grolier, Danbury, CT, 1994). Iraqi death count from the Persian Gulf War cited by ABC National Evening News, October 16, 1994.

53. United States Energy Information Administration, *Annual Energy Review 1993, op. cit.,* pp. 141 and 153.

54. Congress of the United States, Office of Technology Assessment, *Saving Energy in U.S. Transportation, op. cit.,* p. 128.

55. Recent overviews of alternative transportation fuel use include: International Gas Union and the International Association of Natural Gas Vehicles, *Task Force Report: Milan 1994* (Greenaways, England, 1994); and United States Energy Information Administration, *Alternatives to Traditional Transportation Fuels: An Overview* (Washington, DC, June 1994).

56. The Clean Air Act Amendments of 1990, U.S. Public Law #101-594, enacted November 15, 1990.

57. The National Energy Policy Act of 1992, U.S. Public Law #102-486, enacted October 24, 1992.

58. "The 1995 Market for NGVs and Other AFVs," *Natural Gas Fuels,* October 1994; and United States Energy Information Administration, *Alternatives to Traditional Transportation Fuels: An Overview, op. cit.*

59. "Fleet AFV Use Expected to Grow to 4M by 2010," *Oxyfuel News,* June 6, 1994; and "EIA Predicts Increase in Alternative Fuels, Drop in Gasoline Use," *21st Century Fuels,* April 29, 1994. Both reported by *Alternative Energy Network Online Today.*

60. INFORM has studied alternative transportation fuels extensively since 1986. A 1989 INFORM report, *Drive for Clean Air: Natural Gas and Methanol Vehicles,* concluded that natural gas is the best of the leading candidates to play a significantly expanded role as a transportation fuel for the next several decades.

61. United States Energy Information Administration, *Annual Energy Review 1993, op. cit.,* pp. 129, 141, and 189.

62. *Ibid.,* p. 193.

63. "Comparison of Station Prices," *Natural Gas Fuels,* February 1995, p. 13.

64. World Commission on Environment and Development, *Our Common Future* (Oxford University Press, New York, 1987), p. 48.

65. "Books: The Race for Metallic Hydrogen," *The Hydrogen Letter,* July 1994.

66. A. De Jong and H. Van Wechem, "Carbon: Hydrogen Carrier or Disappearing Skeleton," *International Journal of Hydrogen Energy,* Vol. 20, No. 6, June 1995.

67. Solar energy statistics from the Diagram Group, *Comparisons* (St. Martin's Press, New York, 1980), p. 128. Fossil fuel statistics from U.S. Energy Information Administration, *International Energy Annual 1992, op. cit.,* p. xiv. Calculations by INFORM.

68. Sperling, *Future Drive: Electric Vehicles and Sustainable Transportation, op. cit.,* p. 90.

69. R. Williams and G. Terzian, "A Benefit/Cost Analysis of Accelerated Development of Photovoltaic Technology," *op. cit.,* p. 4.

70. D. L. Elliott *et al.,* "An Assessment of the Available Windy Land Area and Wind Energy Potential in the Contiguous United States," (Pacific Northwest Laboratories, Richland, WA, 1991), cited in Alfred Cavallo *et al.,* "Baseload Wind Power from the Great Plains for Major Electricity Demand Centers" (Center for Energy and Environmental Studies, Princeton, NJ, March 1994), p. 6.

71. D. Hall *et al.,* "Biomass for Energy: Supply Prospects," *Renewable Energy: Sources for Fuels and Electricity,* T. Johansson *et al.,* (ed.), (Island Press, Washington, DC, 1993), p. 595.

72. United States Energy Information Administration, *Annual Energy Review 1993, op. cit.,* pp. 169 and 265.

73. "Global Fusion/Hydrogen System is Urged," *The Hydrogen Letter,* July 1994.

74. Calculations from data in United States Energy Information Administration, *International Energy Annual 1992, op. cit.,* pp. xvi and 94.

75. Flavin and Lenssen, *Power Surge..., op. cit.,* p. 33.

76. United States Energy Information Administration, *International Energy Annual 1992, op. cit.,* p. 94.

77. United States Environmental Protection Agency, *National Air Quality and Emissions Report 1993* (Washington, DC, 1994).

78. United States Energy Information Administration, *Annual Energy Review 1993, op. cit.,* p. 229.

79. M. Schiffer, *Taking Charge: The Electric Automobile in America* (Smithsonian Institution Press, Washington, DC, 1994), p. 91.

80. J. O'M Bockris *et al., Solar Hydrogen Energy: The Power to Save the Earth* (Optima, London, 1991), p. 81.

81. DeJong and Van Wechem, "Carbon: Hydrogen Carrier or Disappearing Skeleton," *op. cit.*

82. B. Heydorn, "Hydrogen Industry and Markets," *Proceedings of the First Annual Hydrogen Meeting of the National Hydrogen Association* (Washington, DC, March 8-9, 1990); and "Small Volume Hydrogen Use is Growing," *The Hydrogen Letter,* July 1994.

83. United States Department of Energy, *Hydrogen Program Implementation Plan FY 1994 – FY 1998* (Washington, DC, October 1993), p. 5.

84. DeJong and Van Wechem, "Carbon: Hydrogen Carrier or Disappearing Skeleton," *op. cit.*

85. P. Hoffman, *The Forever Fuel: The Story of Hydrogen* (Westview Press, Boulder, CO, 1981), Chapter 2.

86. When a fuel cell is connected to an electric motor, the unit is an "engine" comparable to the internal combustion engine. An engine, by definition, is a machine that "converts any of various forms of energy into mechanical force and motion." *Webster's New Collegiate Dictionary (G.&C. Merriam Co., Springfield, MA, 1979).*

87. A. Weisman, "Harnessing the Big H," *Los Angeles Times Magazine,* March 19, 1995.

88. Interview with Frank Lynch, Hydrogen Consultants, Inc., May 8, 1995. Hythane is a registered trademark of Hydrogen Consultants, Inc.

Chapter 4

Hydrogen Automotive Engines

Automobiles can be built to run using hydrogen as their sole fuel and producing water droplets as their only emission. The two principal technological paths for hydrogen use in automobiles are:

- Internal combustion engines connected mechanically to conventional vehicles
- Fuel cell engines connected electronically to electric-powered vehicles.

In addition, both combustion engines and fuel cells may be linked with an electrical storage system, such as batteries, to produce another type of vehicle that combines electrical generation and storage technologies in a single vehicle:

- Hydrogen electric hybrid vehicles.

Significant research and development of hydrogen vehicles of all three types is underway around the world today. Internal combustion engines can be adapted relatively easily from burning gasoline to burning pure hydrogen or fuel mixtures containing hydrogen. Internal combustion engine technology is very well developed, having been perfected over the course of a century of use in conventional gasoline-burning automobiles. Until the early 1990s, most research and development of hydrogen vehicles focused on cars powered by internal combustion engines. This research has included vehicles powered solely by hydrogen and other vehicles, called dual-fueled vehicles, that are powered by mixtures of hydrogen and other fuels, including gasoline, diesel, and natural gas.

However, only a small number of prototype hydrogen vehicles powered by internal combustion engines have ever been produced. Worldwide, fewer than 50 are on the road as of mid-1995. Although hydrogen is by far the cleanest of the combustion fuels, burning it releases small amounts of air pollution, most significantly nitrogen oxides. Moreover, hydrogen has always been held back by its high cost and by the high weight and space requirements needed for onboard hydrogen storage, issues discussed in more detail in Chapters 5 and 6. Nevertheless, hydrogen still holds promise as an automotive fuel in conventional internal combustion engines, and a significant effort to research and develop hydrogen combustion engine technology continues.

Fuel cell engines, which offer tremendous efficiency improvements compared with internal combustion engine technologies and are even cleaner, have been successfully demonstrated in a small number of prototype vehicles. In fuel cells, hydrogen and oxygen react chemically, releasing energy directly as electricity, without the heat of combustion. The energy is then used to power an electric motor in an all-electric vehicle. The potential use of hydrogen in advanced fuel-cell-powered electric vehicles has emerged in the 1990s as a new and exciting chapter in automotive history.

Fuel cell engines produce no air pollution, making their environmental promise dramatically cleaner than even that of the clean-burning hydrogen internal combustion engine. Moreover, fuel cell engines are two to three times more energy efficient than today's automobile engines, which are only about 15 percent efficient. This drastically reduces the amount of hydrogen needed to power a fuel cell engine and thus helps to overcome the weight and space drawbacks of hydrogen storage.

Although the United States developed the world's most advanced fuel cell technology in the 1960s, when it began to use hydrogen-powered fuel cell engines in the space program, the world's first hydrogen fuel-cell-powered land vehicle, a bus, was only unveiled in June 1993, in Canada. Within a year, the first fuel cell automobile and bus built in the United States made their debuts. A German fuel cell car was shown publicly for the first time in February 1994. While these advances are important, the low power density of fuel cells remains the key technological problem. Research must provide better ways of increasing the amount of power generated by fuel cells while reducing weight and size. At the same time, costs must drop before fuel cell electric vehicles can emerge as a viable alternative to combustion engine-powered vehicles.

Several hybrid vehicles have also been built and tested. By combining onboard power-generating and electrical storage technologies, hybrids may offer superior vehicle performance compared with vehicles powered solely by any of the single methods of delivering electricity to a motor – combustion engines, fuel cell engines, or electrical storage systems. Two of the principal types of hybrids now under development are both electric vehicles that include electrical storage systems, usually batteries, as part of their propulsion mechanism. In one type, fuel-burning combustion engines power an onboard electrical generator. In the other type, electricity is generated on board by fuel cell engines. Regardless of the way in which it is generated, electricity in hybrid vehicles can be used directly to power an electric motor propelling a vehicle, or it can be stored until needed.

Significant technological issues remain to be addressed before hydrogen engines of any variety can be widely commercialized. Hybrids are more complicated than pure internal combustion engine, battery-powered, or pure fuel-cell systems. However, the pace of technological development of hybrids in the 1990s has been rapid, and these hurdles no longer appear insurmountable.

Hydrogen in Internal Combustion Engines

The physical properties of different transportation fuels require that internal combustion engines be adjusted to suit each fuel, but the basic technology is the same. Hydrogen use in the internal combustion automobile engine dates back to the engine's early development in the 1870s. Claus Otto is generally credited with inventing the spark-ignited internal combustion engine, often called the Otto cycle engine. Otto tested hydrogen, along with just about every other fuel imaginable, in the earliest models of his

A hydrogen-burning internal combustion engine appears very similar to a gasoline-burning engine.
Photo: BMW

engine. Similarly, Rudolph Diesel, inventor of the other major type of internal combustion engine used in automotive applications, tested hydrogen in his compression-ignited diesel engine.

The competition among fuels for automotive applications raged for several decades and was not settled definitively until vast supplies of oil flooded the market early this century. The availability of cheap oil, plus its attributes as an easy-to-handle liquid with a high energy content per unit of volume, sealed the fate of fuels refined from oil – making gasoline and diesel fuel the source of energy for the world's growing automotive fleet.[1]

Mechanics of Hydrogen Combustion Engines

When ignited, hydrogen, like gasoline and other alternative transportation fuels, burns rapidly. When this fuel combustion takes place in a conventional automotive internal combustion engine, the heat generated by breaking the chemical bonds of the hydrogen molecule and the chemical coupling of the hydrogen atoms with oxygen in air explosively expands the remaining air in the engine cylinder, driving a piston downward. The motion of the piston, after several mechanical twists and turns in the drive train, eventually spins the wheels and propels the vehicle. Combustion, the reaction between hydrogen and oxygen atoms in the engine cylinder, converts the chemical energy of hydrogen into thermal energy – that is, heat – which in turn is converted to mechanical energy, motion.

Hydrogen burns well in the internal combustion engine. From an environmental standpoint, hydrogen combustion is especially attractive because it emits only minuscule levels of air pollution. In the absence of a market for hydrogen vehicles, however, engines have not been perfected to burn hydrogen to nearly the extent that the gasoline-burning engine has been fine-tuned over the decades to improve performance. The main struggle for automotive engineers is to refine the technology that will be needed to use hydrogen, once it becomes available and cost-competitive with other automotive fuel alternatives.

History of Hydrogen Use in Combustion Engines

While hydrogen has been tested in laboratory "bench-scale" experiments as a fuel for internal combustion engines since Claus Otto's time, the first serious effort to produce a significant number of hydrogen-powered vehicles was begun by the Germans in the 1920s.[2] Rudolph A. Erren, a German engineer considered the father of the hydrogen vehicle, designed and built such vehicles during a 20-year period ending in the midst of World War II. Erren obtained his first patent for a hydrogen-powered internal combustion automotive engine in 1928. His invention met with approval from the German government, which supported efforts to obtain energy self-sufficiency. At the time, hydrogen produced from plentiful German coal appeared to be an attractive replacement for imported oil.

Within a decade, more than 100 hydrogen-powered trucks were running on German roads near Erren's manufacturing factory in Berlin. At the peak of their use, more than 1,000 cars and trucks were equipped with the Erren engine, including models built by most German automotive manufacturers. Other Erren-sponsored projects included a hydrogen-powered rail car and a submarine used by the German navy during World War II.

UCLA's entry in the 1972 Urban Vehicle Design Contest was a hydrogen-burning hatchback. Photo: Hydrogen Consultants, Inc.

Erren survived the war in Britain and Canada, but his manufacturing plant in Berlin was destroyed by Allied bombings. He returned to Germany in 1945, but his hydrogen projects were never reactivated and hydrogen research lapsed into a period of dormancy.

Several decades later, interest in hydrogen as an automotive fuel was ignited in the United States mainly through the work of Roger E. Billings.[3] In 1966, while a high school student in Provo, Utah, Billings converted a 1931 Model A Ford to run on hydrogen. While in college, Billings converted three other vehicles to run on hydrogen, including a Volkswagen Superbeetle, which competed in the 1972 Urban Vehicle Design Contest, along with another hydrogen vehicle developed at UCLA. Throughout his career, Billings, now the President of the American Academy of Science in Independence, Missouri, has converted 18 vehicles to run on hydrogen. With the exception of his most recent project, which involves a prototype fuel cell vehicle, all have been powered by internal combustion engines.[4]

In the early 1970s, a group of hydrogen enthusiasts in Southern California formed the Perris Smogless Automobile Association. Perris, the small town in which they lived, is located in the eastern, and most polluted, portion of the Los Angeles Basin.[5] Pooling their own limited resources and small donations from government and industry organizations, the Perris group converted several vehicles to run on hydrogen, including a 1950

Studebaker and a Ford pickup truck. Perhaps the most important contribution made by the Perris Association, which dissolved in the mid-1970s, was a series of technical and energy policy statements its members published, which described their work and argued eloquently for the development of pollution-free vehicles using hydrogen.

Momentum for hydrogen vehicle research and development began to move overseas in the late 1970s, primarily to Germany and Japan, as these countries began to look more seriously for alternatives to oil. America's increasingly conservative government spending on alternative energy programs in the early 1980s confirmed this trend. It is only in the 1990s that significant interest in hydrogen vehicles has emerged again in the United States. Only a few American automobiles, including several pickup trucks that are part of two demonstration projects in Southern California, currently burn hydrogen as the sole fuel in an internal combustion engine. About five other demonstration vehicles are powered by fuel mixtures containing hydrogen and natural gas. Chapter 8 discusses in more detail hydrogen vehicle projects in the United States.

Oil vulnerability in the aftermath of the OPEC oil embargo sparked German interest in alternative energy sources, including hydrogen.[6] A 1974 German government report entitled *New Fuels for the Road: Alternative Fuels for Motor Vehicles* identified hydrogen and methanol as two of the most promising long-term alternatives to reliance on oil. The report established research and development priorities for the 1970s, and the German hydrogen research and development effort has continued with varying levels of financial support since then.

Daimler-Benz, the parent company of Mercedes Benz cars, exhibited the first European hydrogen-powered vehicle, a minibus, at the 1975 International Auto Show in Frankfurt, Germany. Several other prototypes followed in the late 1970s. The most significant hydrogen project undertaken and completed in Germany in the 1980s involved the building and testing of 10 Daimler-Benz hydrogen vehicles. The vehicles were test-driven more than 350,000 miles from 1984 to 1988, mostly on the streets of Berlin. The German commitment to hydrogen has expanded in the 1990s. BMW, the second major German auto manufacturer, has also built and road-tested a number of hydrogen-powered vehicles. The German government has invested more than $100 million in these projects and in other hydrogen vehicle-related research, conducted primarily at its hydrogen research center in Stuttgart. Since 1993, German hydrogen research has been reduced because of government fiscal austerity. At the same time, research in Japan, which trailed the Germany effort for a decade, emerged as the world's largest national hydrogen program.

Japan's hydrogen vehicle program, initiated in the early 1970s, was launched as part of Japan's Project Sunshine – a multibillion-dollar effort to develop a wide range of new energy technologies based on renewable resources. The Ministry of International Trade and Industry (MITI) has coordinated Project Sunshine for nearly two decades, with the cooperation of other government agencies, research institutions, and the private sector. Since 1970, scientists at the Musashi Institute of Technology have constructed eight hydrogen-powered automobiles and one hydrogen-fueled truck. In 1991, Mazda unveiled its first hydrogen vehicle, and Nissan and Toyota have projects underway. In 1993, the Japanese government announced a new $2 billion hydrogen vehicle development effort, called the WE-NET program, for World Energy Network, as part of a 28-year, $11 billion expanded "New Sunshine Program."

Several other hydrogen vehicle projects have been initiated in recent years in other countries, especially in Canada and in several countries in Western Europe. Chapter 9 discusses in more detail hydrogen vehicle projects in countries outside the United States.

Challenges Facing Use of Hydrogen in Combustion Engines

In order to burn hydrogen as its sole or primary fuel, conventional internal combustion engine technology requires modification, not major surgery.[7] Engine modifications are needed to respond to several key differences between hydrogen and gasoline. First, as a gas, hydrogen takes up much more volume than an amount of liquid gasoline releasing the same amount of energy upon combustion. Even when compressed to 3,000 pounds per square inch, hydrogen takes up 12 times as much room as gasoline and three times as much room as natural gas. A hydrogen storage system also weighs up to 12 times as much as a comparable gasoline storage tank.

Second, hydrogen ignites very easily. Only 19 Btu of energy are needed to ignite hydrogen, compared with 228 Btu for gasoline and 275 for natural gas. Third, hydrogen can burn in a wide range of concentrations when mixed with air; these concentrations are called flammability limits. The flammability limits range from 4 percent to 75 percent hydrogen mixed with air. A mixture containing any concentration of hydrogen within these limits will ignite. By comparison, gasoline will only burn in concentrations of 1.0 percent to 7.6 percent, and natural gas will burn only in concentrations of 5.3 percent to 15.5 percent. Finally, once ignited, hydrogen burns very quickly, a characteristic measured by the speed with which a flame moves an ignited fuel mixture. The flame speed of hydrogen is about 9 feet per second, about 9 times faster than that of either gasoline or natural gas.[8]

From the standpoint of engine development, there are no technical problems for which remedies remain unknown. The major areas which require additional technological development include:

- Modifying the fuel delivery and engine lubrication systems to accommodate a gaseous fuel that is significantly less energy-dense and cleaner than gasoline.
- Preventing pre-ignition of hydrogen before it enters the combustion cylinder – the principal engine malfunction when hydrogen is used a fuel.
- Obtaining optimal fuel efficiency and power generation through adjustments in the ratio of air to hydrogen in the combustion chamber, turbocharging, and spark timing.
- Redesigning engines to compensate for hydrogen's broad flammability limits, to ensure their safety .
- Adapting engine designs to burn mixtures of hydrogen and natural gas. Fuel blends are an attractive entry route for hydrogen into the automotive fuel market.
- Minimizing the generation of nitrogen oxides without compromising engine performance.

Accommodating a gaseous fuel

Hydrogen is normally delivered to the engine as a gas. While liquid fuels, like gasoline, must be pumped to the engine, gaseous fuel flow is controlled by a pressure regulator between the fuel storage system and the engine. The hydrogen is either mixed with air prior to injection into the engine, a process called external mixture formation, or it is injected directly into the engine, termed fuel injection or internal mixture formation. Most early hydrogen engine development experimented with external mixture formation systems because it was most analogous to the carbureted fuel-mixing technology used in gasoline engines until the mid-1980s. Since then, internal mixture formation systems, closely modeled on fuel injection technology now commonplace in gasoline-burning engines, have become the focus of most hydrogen engine research. Internal mixture formation systems generally offer better solutions to two of the major difficulties encountered in burning hydrogen: pre-ignition and power losses.

Preventing pre-ignition

When a mixture of hydrogen and air enter an internal combustion engine, it has a tendency to ignite prematurely, causing backfiring and potential engine failure. The pre-ignition problem is a result of the low ignition energy of hydrogen. The heat radiating from the walls of the engine's fuel intake system is often sufficient to ignite the hydrogen before it reaches the combustion chamber.

A number of options have been developed to prevent pre-ignition in external fuel mixing systems. A common approach is to inject small amounts of water vapor to cool the fuel mixture. Alternatively, exhaust gases can be recirculated through the engine. Either option lowers the temperature of the fuel mixture and inhibits spontaneous pre-ignition.

Internal fuel mixture technology greatly lessens the pre-ignition problem compared with external fuel formation because it separates hydrogen and air until they are actually injected into the cylinder. Because hydrogen is not in contact with air prior to injection into the engine, no combustion can occur. Precise timing of fuel injection just before the spark plug fires can effectively prevent pre-ignition in internal fuel mixing systems.

Fuel efficiency and power generation

Because hydrogen is a gas with far less energy per unit volume than gasoline, large volumes of hydrogen must be injected into a combustion chamber to deliver the same amount of energy as gasoline. Hydrogen displaces about 29 percent of the air in the combustion chamber of an internal combustion engine, while gasoline with the same energy content displaces less than 2 percent.[9] Because of this, hydrogen engines require larger fuel-mixing equipment and fuel injectors with wider apertures than are part of gasoline fuel-delivery systems. Moreover, the large displacement of air by the hydrogen causes greater power losses compared with gasoline engines. It is the force of the rapidly expanding heated air in the combustion chamber that drives the engine piston downward after fuel ignition. If more space is consumed by the fuel, there is less room for air in the chamber; consequently, less force is exerted on the piston and less power is generated by the engine. Power losses of nearly 30 percent have been experienced when conventional internal combustion engines have been converted to run on hydrogen.[10]

However, these problems can be overcome; the same technological options exist to restore the power to a hydrogen engine as have been developed for natural gas engines, where power losses of 20 percent were experienced in the first converted vehicles. For example, the compression ratio of the engine can be increased (to exert more force on the piston), a turbocharger can be inserted to force more air into the combustion chamber, and the air/fuel mixture can be cooled prior to ignition to increase its density. Indeed, by using these techniques, hydrogen test vehicles have demonstrated power increases up to 25 percent greater than their gasoline-powered counterparts.[11]

Flammability

A final physical characteristic of hydrogen that must be addressed by redesigning conventional engine technology is the ability of hydrogen to burn in widely varying concentrations of air.

The wide flammability limits of hydrogen, combined with its very low ignition energy, raise one of the greatest safety concerns during the production and handling of hydrogen: unwanted combustion (see Chapter 7).

But hydrogen's wide flammability has its advantages as well; in the engine's combustion cylinder, wide flammability means that only very small amounts of fuel need to be injected into the engine in order to keep the engine running. The ability of hydrogen to burn in the presence of excess quantities of air, termed "lean burn" conditions, promotes complete fuel combustion, greater fuel efficiency, and cooler engine operating environments, which minimizes the formation of nitrogen oxide pollution. Fuel efficiency is also enhanced because, as a gas, hydrogen mixes more thoroughly than liquid fuels with air prior to combustion. This promotes complete and even combustion of the fuel when ignition takes place. Research to date indicates that, by operating under lean burn conditions, the hydrogen engine's thermal efficiency, a measure of fuel economy similar to miles driven per gallon of gasoline, increases an average of 22 percent compared with engines running on gasoline.[12]

Dual-fuel engines lessen combustion problems

Most hydrogen vehicle research to date has focused on engines that replace gasoline or diesel fuel with hydrogen. Alternative engine designs – called dual-fuel engines – have also been developed to burn mixtures of fuels – including gasoline, diesel, natural gas, and hydrogen. The use of dual-fuel engines may offer a practical step on the pathway to a transportation system fueled by hydrogen in the future.

Most dual-fuel hydrogen research has focused on fuel mixtures containing only a small percentage of hydrogen. Mixing small amounts of hydrogen in gasoline, a process called hydrogen enrichment, was first proposed by scientists at the Jet Propulsion Laboratory in California in the mid-1970s as a method to reduce nitrogen oxide emissions from gasoline burning.[13]

Since the late 1980s, interest has been rising in the use of mixtures of natural gas and hydrogen in engines that have already been modified to burn natural gas. Several research projects in Colorado, California, and Pennsylvania have tested the performance of vehicles running on a mixture of 95 percent natural gas and 5 percent hydrogen by energy content, coined as Hythane by the principal developer, Hydrogen Consultants, Inc. Results indicate that addition of just 5 percent hydrogen can reduce hydrocarbon and carbon monoxide pollution by up to 50 percent compared with the already low levels of pollution achieved by burning pure natural gas.[14] Other tests at the Florida Solar Energy Center using fuel mixtures containing up to 15 percent hydrogen by energy content suggest that even greater pollution reductions may be possible in dual-fuel hydrogen engines, reaching as far as 80 percent below the California ultra-low emission standards for these pollutants.[15]

In addition to the dramatic air quality benefits, there are several attractive attributes of dual-fuel engines that use hydrogen. First, the wide flammability limits, low ignition temperature, and high speed with which hydrogen burns, if properly regulated in dual-fuel engines, can improve overall engine performance. Second, the use of hydrogen, if generated from a renewable energy resource, replaces a portion of the dominant fuel, reducing fossil fuel use. This helps achieve energy diversification that can be beneficial in areas where conventional fuels are in short supply or are very expensive.

Perhaps the most appealing aspect of dual-fuel technology is that it may provide a viable pathway for the gradual introduction of hydrogen as a transportation fuel. Research to date indicates that no major engine modifications are necessary in order to burn mixtures containing hydrogen in concentrations less than about 20 percent.[16] By leveraging small quantities of hydrogen into a transportation system that is still dominated by gasoline and diesel fuels or by widely used alternatives such as natural gas, dual-fuel engines allow hydrogen to gain a foothold – without the costs and efforts involved in building vehicles specially designed to burn hydrogen. Dual-fuel engines also lessen one of the major problems in switching to hydrogen – the lack of a refueling infrastructure – because dual-fuel vehicles are usually capable of operating solely on the dominant fuel when hydrogen is not available.

On the other hand, the environmental benefits of hydrogen are not maximized when it is used only as a small portion of a vehicle's fuel mix. Furthermore, although dual-fuel engines can displace some fossil fuel demand, they still depend largely on them, thus failing to achieve the objective of basing transportation on sustainable energy resources.

Environmental Factors

The sole by-product of hydrogen combustion is water. The only visible emission from the tailpipe of a hydrogen-powered vehicle is a wisp of steam and an occasional droplet of water. Side reactions not involving hydrogen can produce some levels of other air pollutants – hydrocarbon and nitrogen oxide emissions – but these can be contained to such low concentrations that the exhausts from some hydrogen test vehicles have been shown to be cleaner than the air sucked into the engine intake. In these circumstances, the engine acts as a "minus-emission" machine because the air pollutants in the ambient air sucked into the engine decompose as the hydrogen burns.[17]

Less air pollution than fossil fuels

In terms of air pollution, hydrogen is clearly superior to any fossil fuel when burned in an internal combustion engine. Hydrogen molecules contain no carbon, so no carbon dioxide is produced when hydrogen is burned. Hydrogen burning, therefore, does not contribute to the buildup of carbon dioxide in the atmosphere, as long as the method of producing the hydrogen itself, discussed in Chapter 6, does not release carbon dioxide, either.

Carbon monoxide, hydrocarbon, and particulate pollution, including toxic air pollutants in automotive exhausts, result from the incomplete combustion of molecules containing carbon. Bereft of carbon, burning hydrogen produces none of these pollutants either. However, because internal combustion engines are lubricated by oil, minuscule emissions result from the burning of engine oil coating the walls of the combustion chamber. Carbon monoxide, hydrocarbon, and particulate emissions from hydrogen test vehicles operating without any pollution-control equipment are generally less than 10 percent of the emissions from a comparable gasoline-powered vehicle operating with advanced pollution-controlling catalytic convertors, and are often far less than 10 percent. Addition of control technology, if deemed necessary, would essentially eliminate emissions of these pollutants.[18]

Nitrogen oxides: The major environmental challenge

Nitrogen oxide emissions pose the main environmental issue of concern for hydrogen combustion. Nitrogen oxides are ozone precursors; that is, they contribute to the buildup of ozone in the lower atmosphere, where it becomes a respiratory irritant and can damage plant life. Nitrogen oxides also cause respiratory problems on their own and are a key component in acid rain.

As with hydrocarbons, no nitrogen oxides are produced directly from hydrogen combustion. The heat generated by the combustion, however, is sufficient to cause a chemical reaction between oxygen and nitrogen in the air in the combustion chamber (ambient air is composed roughly of 20 percent oxygen and 80 percent nitrogen), fusing them into nitrogen oxide compounds. This chemical reaction occurs at temperatures above 2,700 degrees Fahrenheit; since the flame temperature of hydrogen is about 4,480 degrees Fahrenheit, roughly the same as gasoline, the combustion of either fuel can produce nitrogen oxides.

Although the generation of nitrogen oxides is an issue that must be addressed by hydrogen vehicles powered by internal combustion engines, its control is less complicated than for gasoline-powered vehicles. Engine designs that include exhaust gas recirculation, fuel cooling, lean-burn operating conditions, or water injection reduce the generation of nitrogen oxides to levels below the current federal automotive pollution standard, without use of specialized pollution control equipment. If additional pollution controls are necessary or desirable, catalytic convertors for hydrogen vehicles could be specially designed to achieve a very high degree of nitrogen oxide control.

Water: Not a significant concern

The production of water by hydrogen vehicles raises a final environmental question that is not a public health issue but which must be addressed scientifically nonetheless. Could a proliferation of hydrogen vehicles produce enough water to alter global or regional weather patterns? Research to date indicates that this is not a concern at the global level and probably not a significant cause of concern at the regional or local level. However, more study is needed to address this question fully, especially regarding potential microclimate effects.

In cases where hydrogen is originally produced by splitting water molecules, the generation of water in a hydrogen vehicle simply completes a cycle that does not alter the total global water balance. In cases where hydrogen is produced from fossil fuels, for example, by steam reforming of natural gas, the amount of water produced is about twice the amount produced by burning the fossil fuel directly. In fact, the burning of fossil fuels produces a lot of water. The hydrogen contained in the molecular structures of all fossil fuels combines with oxygen as the fuels burn to produce water. Each year, 7 billion tons of water are added to the atmosphere from the burning of fossil fuels. This could double to 14 billion tons if hydrogen produced from fossil fuels replaced every direct use of fossil fuels.[19]

The amount of water involved under any scenario, however, is small compared with the 14 trillion tons of water vapor that are held in the atmosphere. Each year, 551 trillion tons of water is cycled between the atmosphere and the ground through rainfall and evaporation. One study has calculated that even if the entire world's usage of fossil fuels was replaced by hydrogen produced from sources other than water, the total amount of water involved in the hydrogen cycle from production through the tailpipe would equal 0.003 percent of global water evaporation.[20]

From a regional perspective, however, water production from hydrogen combustion might in unusual circumstances affect local weather conditions. A small increase in local average humidity levels could increase dew, fog, and the icing of roads. Even these effects are likely to be less than the impacts that currently result from the massive steam evaporation from fossil-fuel-burning power plants.[21]

Hydrogen Fuel Cell Engines

The fuel cell was discovered in England by Sir William Grove when he reasoned, and proved experimentally in 1839, that since electricity could be used to split the water molecule into hydrogen and oxygen in a device called an electrolyzer, it must also be scientifically feasible to run an electrolyzer backward to generate electricity and water from hydrogen and oxygen.[22] Such a device is called a fuel cell. Throughout the late 19th and early 20th centuries, scientists continued to experiment with various rudimentary fuel cell designs. Today, fuel cells are being used to power vehicles in demonstration projects around their world, but fuel cell engines require additional refining before they become a commercially viable automotive technology – possibly within the next decade.

A fuel cell engine, at first blush, appears to be magic. Having no moving parts, the fuel cell is silent, produces no pollution and little heat, and provides up to three times more energy per unit of hydrogen consumed than an internal combustion engine. Although not magic to scientists who understand the electrochemical principles involved, fuel-cell-powered transportation systems have long remained only a pipedream because of the difficulties in attaining the high performance required of automotive engines under harsh operating environments. That is, until 1993, when the world's first fuel-cell-powered vehicles were built. Although still years away from commercialization in automotive applications, hydrogen fuel cell engines are now among the newest and most promising prospects to replace internal combustion engines.[23]

Because hydrogen fuel cell engines generate electricity, fuel-cell-powered vehicles are a type of electric vehicle, similar to vehicles powered by electricity stored on board in batteries or in other storage devices, such as flywheels or ultracapacitors (discussed in Chapter 5).

Mechanics of Fuel Cell Engines

Individual fuel cell units consist of two electrodes imbedded or submerged in an electrolyte, and a wire connecting the electrodes without passing through the electrolyte. Each unit is capable of generating slightly less than one volt of electrical power. To obtain the power required to propel an automotive requires that hundreds of fuel cell units be connected in a series of generating units called "stacks."

By analogy, in the internal combustion engine the reaction between hydrogen and oxygen is initiated by a spark and the energy released by the reaction appears as thermal energy, called the heat of combustion. The fuel cell reaction is sometimes referred to as "cold combustion" because the

Fuel Cells: Electrolysis in Reverse

Electrolysis of water, or the splitting of water into its atomic constituents hydrogen and oxygen, is a simple high school science class experiment. Wires are connected to the positive and negative terminals of the source of electrical current. The loose ends of the electrical wires are connected to plates of metal, called electrodes, submerged in a jar of salty water, called the electrolytic solution or electrolyte. An electrolyte is a nonmetallic substance that permits the movement of charged particles, called ions. In a common car battery, for example, sulfuric acid is the electrolyte and lead is the electrode.

The current flows in a circle from the power source through the water and back to the power source. As this occurs, bubbles of gas form at the electrodes. The bubbles at one negatively charged electrode, called the anode, are pure hydrogen. At the other, positively charged electrode, called the cathode, pure oxygen is produced.

In electrolysis, the electrical energy from the power source does the work of splitting the water molecule. The electrical energy is sufficient to overcome the energy of the chemical bonds holding hydrogen to oxygen in water molecules. It does this without moving parts and with the generation of only a small amount of heat. Electrolysis of water is currently the most viable way to produce hydrogen when using electricity as the energy source. The technology is quite advanced, and energy conversion efficiencies above 80 percent are commonplace; conversion efficiencies concern the relationship between energy in the hydrogen molecules and the energy in the electricity expended to produce them.

A fuel cell, in essence, is an electrolysis unit that operates in reverse. Instead of electrical energy forcing hydrogen and oxygen in a water molecule apart, hydrogen and oxygen merge in a fuel cell – forming water and releasing energy as electricity. In a common type of fuel cell, anode electrodes, composed of catalysts that facilitate chemical reactions, break down hydrogen gas into positively charged hydrogen ions and negatively charged electrons. The hydrogen ions travel through an electrolyte to the cathode, where they meet oxygen. The electrons are prevented from traveling through the electrolyte, which is impervious to them. Instead, electrons are forced to move to the cathode through a wire. This movement of electrons is an electrical current that can be tapped to perform work – for example to power an electric motor. The hydrogen ions, electrons, and oxygen react chemically at the cathode to form water.[24]

Figure 1. How a Fuel Cell Works

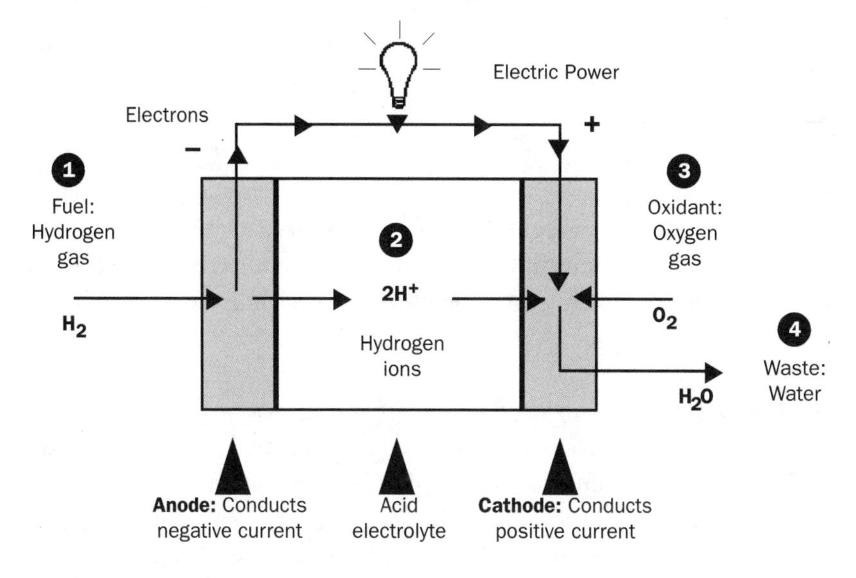

1. Hydrogen gas is piped to the anode and passes through an acid electrolyte – or in the case of Energy Partners' version, a permeable plastic membrane.
2. In this reaction, the hydrogen molecule splits, giving off a stream of electrons, or electricity, and creating two hydrogen ions with a positive charge.
3. On the other end of the cell, oxygen gas is piped through the cathode.
4. When exposed to electric current, the oxygen combines with the hydrogen ions, forming water.

Source: Energy Partners as published in *The New York Times.*

energy does not appear as heat, but rather as electricity.

Fuel cells are also analogous to electric batteries. Both are electrochemical devices that convert energy released by chemical reactions directly into electricity. The difference lies in the source of the chemical energy. Batteries store all the chemicals within the battery itself. As the battery operates, the chemicals react and eventually are used up. Such "dead" batteries must be recharged, a process that uses electricity from an external source to reverse the chemical reaction and restore the reactants. Thus, batteries are a means to store a limited amount of electricity temporarily in chemicals until it is needed. In fuel cells, on the other hand, the chemical reactants, hydrogen and oxygen, are continuously supplied from a storage system distinct from the fuel cell. As long as the fuel supply continues, fuel cells are capable of generating electricity indefinitely. Thus, fuel cells are engines that generate, rather than store, electricity.

Once electricity is generated in a fuel cell, the remainder of the vehicle is similar to any electric vehicle. The direct current (dc) produced by the fuel cell must be converted to alternating current (ac) by a dc-ac inverter to complement most electric vehicle designs. An electric motor then converts the ac electricity into mechanical motion. Elaborate electronic controls are needed to regulate the flow of electricity to the wheels as driving patterns change.

Greater energy efficiency than combustion engines

Capturing the energy produced by the reaction between hydrogen and oxygen directly as electricity in a fuel cell offers tremendous advantages for powering automobiles compared with its release as heat in internal combustion engines. The energy efficiency of the current generation of fuel cells – the percent of energy contained in hydrogen that is used or delivered as movement – is two to three times as great as the efficiency of converting the energy of fuels burned in internal combustion engines. Most of the heat generated by the combustion of hydrogen in an internal combustion engine is lost to the engine walls or radiator and dissipated into the air as thermal pollution or friction; less than 20 percent is normally translated into the mechanical motion of the piston.[25]

A fuel cell, on the other hand, has no moving parts to create frictional losses; therefore, a much higher percentage of the reaction energy of hydrogen and oxygen is captured by electricity, which can be put to work propelling the car. The theoretical, or scientifically possible, maximum efficiency of a fuel cell is 85 percent[26]; the practical maximum efficiency expected for fuel cell technologies currently under development is between 40 percent and 50 percent.[27]

Zero-emission vehicles

A fuel cell has all the environmental advantages of a hydrogen-powered internal combustion engine, plus a few more. In both systems, the only chemical by-product from the use of hydrogen is water. The fuel cell engine does not require the use of lubricating oil, however. Thus, the small emissions of hydrocarbons and carbon monoxide from the burning of lubricating oils that leak into internal combustion engines is eliminated in fuel cells. More important, because no combustion occurs in a fuel cell, fuel cells do not mix oxygen and nitrogen at high temperatures; thus no nitrogen oxide emissions are formed and little waste heat is produced. From a standpoint of air pollution, fuel cell vehicles using hydrogen are truly zero-emission vehicles.

History of Fuel Cell Technology: Spurred by the Space Race

The first fuel cell capable of generating a significant amount of electricity was not built until more than a century after the discovery of the electrochemical principals of the technology. The British scientist Francis Bacon manufactured a system in 1932 that consisted of individual fuel cells connected in series, so that the total flow of current was equal to the sum of the electricity generated by each individual cell. By stacking cells, Bacon was able in 1959, after more than a quarter of a century of effort, to produce a unit capable of generating 5 kilowatts of electricity, enough to serve the electrical needs of about two homes.

Fuel cell technology might have remained a scientific curiosity, however, were it was not for the "space race" in the 1960s. Scientists discovered that fuel cells are the perfect power supply for space travel because their fuel, hydrogen, is the lightest element in the universe. The task of getting the fuel from earth into space, therefore, is the easiest and cheapest. Because fuel cells produce electricity whenever needed, they are an ideal complement to solar panels (which also generate electricity, but only when exposed to the sun), assuring a continuous supply of electricity aboard spacecraft.

In the early 1960s, the National Aeronautics and Space Administration (NASA) invested more than $20 million in the development of hydrogen-powered fuel cells for use in the space program. This started a process that resulted in the worldwide leadership of the United States in fuel cell technology, although terrestrial applications remain limited. Working with generous NASA funding, scientists were able to increase the lifespan of fuel cells from 100 to thousands of hours, reduce system weight by 90 percent, and slash production costs. The expense of fuel cells for spacecraft is still extremely high, about 100 times more expensive than conventional electrical generation technologies used on Earth. By 1965, fuel cells were standard equipment aboard spacecraft, a role that continues today.[28]

Challenges for Hydrogen Fuel Cell Engines

Although it is clear that fuel cells have helped take us to the moon, it may take a decade before they are practical and cheap enough to take us to the local store. The same production, storage, and safety issues confronting hydrogen use in internal combustion engines also face hydrogen when used in fuel-cell-powered vehicles. These issues are less troublesome when hydrogen is destined for use in fuel cells, however, because high fuel cell efficiencies mean that less hydrogen is needed.

The technological issues facing fuel cell engine technology, however, are much more serious. These technological hurdles must be overcome before fuel cells can effectively compete with internal combustion engines in automotive applications. And even if the performance, reliability, and durability of fuel cells could match internal combustion engines, the cost of fuel cells must be reduced if they are to become a practical energy technology for personal transportation.

The key technological issues facing fuel cell engines are summarized below:

- **Weight** An internal combustion engine weighs about 2 to 4 pounds for every kilowatt of power it generates. Fuel cells today are more than 100 times more powerful than in 1961, per area of fuel cell membrane surface.[29] Nonetheless, the current generation of fuel cells is still heavier than their combustion engine counterparts, weighing between 15 and 30 pounds per kilowatt of power. Although the higher efficiencies of fuel cells in large part compensate for their heavier weight – meaning that smaller engines can deliver more power to the wheels – the weight of fuel cells must be reduced further. Recent developments in proton-exchange membrane research indicate that a fuel cell with a weight as low as 4 pounds per kilowatt may be demonstrated within the next few years. Fuel cells of the future could potentially weigh less than half the weight of combustion engines in use today.

- **Catalyst Performance** Reducing the amount of rare and expensive catalysts used in fuel cell electrodes without cutting power output poses a major technological challenge. Recent developments in thin-plate electrode technology for proton-exchange membrane fuel cells has reduced the amount of platinum needed to nearly as much as is the platinum currently contained in the catalytic converters of gasoline-burning vehicles. The cost of platinum electrodes for a fuel cell large enough to power an automobile has dropped from more than $30,000 in 1984 to about $500 today.[30] Catalysts containing alloys that do not include any rare metals are another promising approach to improving catalyst performance, while reducing costs and fuel cell weight.

- **Durability** Fuel cells, like rechargeable electric batteries, tend to degrade over time and eventually lose their ability to generate electricity. The lifetime of an internal combustion engine often exceeds 5,000 hours of operation. In comparison, the earliest fuel cells lasted only hours. Although some few cells have operated for more than 10,000 hours, most fuel cells today operate for a few thousand hours, and durability remains an issue.

- **Contamination** Just about any liquid or gaseous fuel burns in an internal combustion engine or passes through, leaving the engine unscathed. On the other hand, most fuel cells, particularly those that operate at low temperatures, are quickly deactivated if exposed to even minute concentrations of gases such as carbon monoxide or carbon dioxide. Contamination-proof or self-cleaning catalysts are a possible solution.

- **Temperature** The performance of all fuel cell technologies is temperature-sensitive. Even the proton-exchange membrane fuel cells, which operate at low temperatures, require that optimal operating temperatures be reached quickly and maintained while the engine is running. Furthermore, materials need to be developed that can survive the high operating temperatures required by the molten carbonate and solid oxide fuel cell technologies.
- **Water Control** Fuel cells produce water as they operate. If water accumulates in a fuel cell, it dilutes the reactants – reducing and eventually stopping the power output of the engine. On the other hand, fuel cells need to be kept moist in order to operate. Thus, it is important to regulate the water level in a fuel cell by draining excess water as it is produced.
- **Engine Starts and Variations** Automobiles must be able to start at the turn of a key. Some types of fuel cells require lengthy warm-up periods before the engines generate power sufficient to move a vehicle. Furthermore, the output of some fuel cells cannot be quickly changed in response to the multiple variations in speed and acceleration normal to driving patterns. Although the issue is not a major problem with proton-exchange membrane fuel cells, the ability of other types of fuel cells to operate flexibly over changing operating conditions needs to be enhanced.
- **Fuel Flexibility** Fuel cells use hydrogen and oxygen to generate water and electricity. Fuel cells that can obtain oxygen by extracting it from air are simpler and cheaper to operate than fuel cells that require pure oxygen. Similarly, fuel cell technology that can obtain hydrogen by extracting it from other, easier-to-store fuels such as natural gas or methanol may hold cost and operation advantages compared with fuel cells that can use only pure hydrogen.
- **Cost** Practical solutions to the above challenges will help lower the cost of fuel cells. A gasoline-fueled internal combustion engine now costs $50 per kilowatt of power (one kilowatt is roughly equal to 1.2 horsepower). Fuel cells now cost at least 20 times as much, or a

minimum of $1,500 per kilowatt. The high cost of fuel cells is due partly to the high price of electrode materials, particularly platinum, and the fact that each fuel cell is virtually handmade. Fuel cell units used in the world's first prototype vehicles cost more than $100,000. (The life-cycle cost of hydrogen fuel cell vehicles are discussed further in Chapter 7.)

Five Fuel Cell Technologies

There are five basic fuel cell technologies under development, each with a distinct performance profile. The applicability of these technologies for use in vehicles, and the relative potential of each technology, depends on a number of factors, including the "challenges" outlined in the previous section: cost, weight, size, durability, ease of contamination, temperature and water requirements, engine starts and variations, and fuel flexibility.

The five technologies are discussed below in the order of their potential for practical application in vehicles: proton-exchange membrane; phosphoric acid; alkaline; molten carbonate, and solid oxide fuel cells. (See Table 1 on page 119 for a summary of these techologies.)

Proton-exchange membrane fuel cells

Proton-exchange membrane fuel cells are the greatest hope of fuel cell advocates as the technology most likely to unseat the internal combustion engine, possibly sometime early next decade. The low operating temperature, durability, compact size, adjustable power outputs, and quick-starting characteristics of proton-exchange membrane fuel cells make them ideal candidates for use in vehicles and the most effective technology in addressing the problems of fuel cells outlined above.

The first non-space vehicle to use a proton-exchange membrane fuel cell was a two-person research submarine built in Canada in 1989 and equipped with a 2-kilowatt fuel cell. It performed 16 test dives off the coast of British Columbia with favorable results. Proton-exchange membrane fuel cells were also the engine of choice for two of the world's first fuel cell land vehicles, the "Green Car" built in 1993 by Energy Partners, Inc., in Florida and the Ballard fuel cell bus, which began road tests in

Energy Partners' Green Car carries stacks of proton-exchange membrane fuel cells. Photo: Energy Partners.

Vancouver, Canada in 1993. In early 1994, Daimler-Benz began test driving its version of a proton-exchange membrane fuel-cell-powered automobile.

Non-transportation applications of proton-exchange membrane fuel cells are also proliferating. A 25-watt version is being marketed as a replacement for nickel-cadmium batteries in video cameras. Other applications include substitutes for batteries in portable laptop computers, in smoke detectors, and in electronic equipment carried by infantry soldiers.

Proton-exchange membrane fuel cells owe their existence to the United States space program. The cells' viability is a result of an ion-exchange membrane first developed by General Electric in the early 1960s. The membrane, which serves as an electrolyte, is impermeable to electrons while freely permitting positively charged hydrogen ions to pass. The General Electric membrane, made primarily of polystyrene, was used in the proton-exchange membranes that, beginning in 1964, powered the Gemini series of spaceships in planetary orbits. In 1968, DuPont invented a superior membrane, called Nafion, which has been the choice of proton-exchange membrane fuel cell manufacturers since then.[31]

Nafion is composed primarily of perfluorinated Teflon, which is extremely sturdy and chemically stable for long periods under harsh operating conditions. The durability of Nafion membranes is one advantage proton-exchange membrane fuel cells hold over cells based on other

Table 1. Five Fuel Cell Technologies

(Presented in order of their potential for application in vehicles)

Fuel Cell	Advantages	Operating Restrictions	Applications	Maximum Efficiency*
Proton-exchange membrane	•Durable •Doesn't require pure oxygen •Operates at moderate temperature (150°F) •Lightweight, compact •Starts quickly •Operates over range of power outputs	•CO at concentrations of of 0.5% deactivates membrane •Requires saturation in H_2O	•Daimler-Benz vehicles •Energy Partners' "Green" car •Ballard bus •Gemini spacecraft •Research submarine (Canada)	60%
Phosphoric Acid	•Tolerates CO_2	•Cannot operate if CO concentrations exceed 1% •High operating temperature (between 200 and 350°F) •Degrades at partial power output •High cost	•DOE-funded development of urban buses (DC) •Stationary projects (Los Angeles, Japan)	40%
Alkaline	•Starts quickly •Lightweight •Operates at room temperature and pressure	•Requires pure oxygen (can't tolerate CO_2) •High cost	•Apollo moon units •Proposed bus project (Belgium)	70%
Molten Carbonate	•Tolerates ambient air •Can also use natural gas or methanol	•Extremely high operating temperature (1200°F)	•Small stationary unit (California)	>60%
Solid Oxide	•Tolerates ambient air •Can also use natural gas or methanol	•Extremely high operating temperature (above 1900°F)	•Small stationary unit (Japan)	>60%

* Efficiency is the percent of energy contained in the hydrogen fuel that is delivered as electricity.

electrolytes. The estimated life of the best proton-exchange fuel cells today is 150,000 to 200,000 miles, which is a slight improvement over the average internal combustion engine and vastly better than the 10,000- to 30,000-mile lifespan of most automotive batteries now under development.

Perhaps the best news offered by proton-exchange membranes is that they operate at moderate temperatures, about 150 degrees Fahrenheit, thereby eliminating the need in several other fuel cell technologies to attain and maintain higher temperatures. A related and equally important benefit is that proton-exchange membrane fuel cells can be started very quickly, in part because there is no need to wait for the fuel cell to be heated in order to initiate the flow of electricity. Furthermore, proton-exchange membrane fuel cells do not require pure oxygen; they can operate on ambient air. They are also relatively light and compact, weighing less than half the weight of a comparably sized phosphoric acid fuel cell. Finally, proton-exchange membrane fuel cells can be operated over a wide and rapidly changing range of power outputs.

Proton-exchange membranes are not without problems. They are easily rendered inoperable by carbon monoxide contamination: Concentrations as low as 0.5 percent in the air intake quickly deactivate proton-exchange membranes. Thus, scrubbing carbon monoxide from the air intake may be required to ensure the durability of proton-exchange membranes. This is easier and cheaper to accomplish than to run the fuel cell with pure oxygen, but it still presents technical challenges. Another concern is that membranes must be constantly saturated in water to prevent drying, loss of efficiency, and their eventual destruction from cracking. This is usually accomplished by pressuring the system to ensure that even the inner reaches of the membrane are continuously bathed in water.

Phosphoric acid fuel cells

Three characteristics of phosphoric acid fuel cells – large size, high operating temperature, and the degradation of materials when idling or operating at slow speeds – are drawback to their use in vehicles. Nonetheless, their high status of development and their lower price relative to other fuel cell options has captured the attention of the United States Department of Energy. The DOE is currently funding a program to develop a phosphoric acid fuel cell for use in urban buses. The project, discussed in Chapter 8, includes field-testing of three buses. The first bus began operating in Washington, DC, in April 1994. It represents the world's first successful effort to operate a bus powered by a phosphoric acid fuel cell.

Phosphoric acid fuel cells are now being commercialized mainly for stationary applications such as supplemental power generators in polluted areas. A 200-kilowatt model began operating in Los Angeles in 1992. Built at a cost of $500,000, it provides electricity and hot water to the new headquarters of the region's chief environmental protection agency, the South Coast Air Quality Management District. More than 50 other identical units are operating or on order elsewhere, including 10 in the Los Angeles area and one in Sacramento. In Japan, 18 even larger units are grouped together to form one phosphoric acid fuel cell complex at a power plant that is capable of generating 11,000 kilowatts of electricity.

The electrodes of phosphoric acid fuel cells are made of fine particles of platinum in layers only 100 microns thick – about the thickness of a human hair. The platinum is fused onto sheets of carbon that are separated by a graphite plate suspended in silicon carbide soaked in a solution of phosphoric acid. Like proton-exchange membranes, phosphoric acid fuel cells have the advantage of being able to operate using air as opposed to pure oxygen. Unfortunately, phosphoric acid fuel cells, also like proton-exchange membrane fuel cells, are easily ruined when exposed to carbon monoxide at concentrations above 1.0 percent. Therefore, air must first be thoroughly cleaned of all but small traces of carbon monoxide before injection into a fuel cell.

Another concern is that phosphoric acid fuel cells do not work at temperatures below 200 degrees Fahrenheit; in practice, temperatures above 350 degrees are needed to produce the fastest reaction rates. As a result, the start-up of a phosphoric acid fuel cell is slow until adequate operating temperature within the cell is reached. Also, the carbon sheets in phosphoric acid fuel cells tend to degrade quickly when the cells operate at partial output – for example, when a vehicle is idling. Their best performance is achieved only when the cells operate continuously at full power. The average efficiency achieved with phosphoric acid fuel cells is only about 40 percent, the lowest among the five technologies now under development.

Finally, the current generation of phosphoric acid fuel cells are the bulkiest of the five fuel cell systems.

Alkaline fuel cells

Alkaline fuel cells are the oldest fuel cell technology – first built by Francis Bacon in the 1930s. Alkaline fuel cells powered the early spacecraft, including the Apollo capsules that first journeyed to the moon. The Apollo fuel cell operated at about 500 degrees Fahrenheit, weighed 245 pounds, and produced only a little more than 1 kilowatt of power, barely enough to power a handful of lightbulbs on Earth. NASA still uses alkaline fuel cells; four 15-kilowatt units are aboard each space shuttle. To date, their requirement for pure oxygen and their high cost have limited interest in developing alkaline fuel cells for automotive applications: Alkaline fuel cells cost about $100,000 per kilowatt for the space program versions, compared with $800 to $1,500 per kilowatt for a natural gas or coal electrical-generating station on Earth. However, one bus project underway in Belgium, the Eureka Bus (discussed in Chapter 9), uses an alkaline fuel cell.

Alkaline fuel cell design is the most similar to an electrolysis system operating in reverse; in fact, the alkaline fuel cells used in the space program are capable of operating in either direction – making electricity from hydrogen and oxygen or using electricity to regenerate hydrogen and oxygen from water. Electrodes, typically made out of nickel or silver, are submerged in an alkaline solution, most commonly containing potassium hydroxide.

The advantages of alkaline fuel cells are that they start quickly, can use cheaper electrodes than those made with platinum, and are lighter than other fuel cell types.[32] They work well at atmospheric pressure and room temperatures, although efficiencies are improved at higher pressure and temperatures. Efficiencies near 70 percent have been achieved, the highest level of any fuel cell technology.

On the other hand, alkaline fuel cells require pure oxygen and become inoperable if ambient air is substituted because carbon dioxide in the air deactivates the electrodes. This is part of the reason for the high cost of alkaline fuel cells and for the reliability problems encountered if even very low impurity concentrations appear in the fuel supply.

Molten carbonate and solid oxide fuel cells

Molten carbonate and solid oxide fuel cells are the two remaining types of fuel cell technologies that are attracting increased attention as potential replacements for the internal combustion engine. Both hold promise, but are only in the early development stages. No vehicles, including spacecraft, have yet been built equipped with either technology.

Molten carbonate and solid oxide fuel cells are currently being tested in a few stationary source applications. The most technologically advanced solid oxide fuel cell today is a 20-kilowatt unit built by Westinghouse and operating in Japan. A 70-kilowatt molten carbonate fuel cell developed by Energy Research Corp., was tested at the company's facility in Connecticut in 1994. A 2-megawatt molten-carbonate fuel cell test unit is being built in Santa Clara, California and is scheduled to begin operation by early 1996. A 2,000-kilowatt molten carbonate system has been proposed for construction by 1996 to generate electricity at a medical complex in California. A number of other high-temperature fuel cell designs are being tested around the world, but most are smaller than 10 kilowatts in capacity.

In the molten carbonate fuel cell, the electrolyte is usually a mixture of lithium and potassium carbonate solutions. The cathode is normally nickel oxide and the anode is usually an alloy of nickel. The carbonate electrolytes are heated to 1,200 degrees Fahrenheit, at which temperature they become molten fluids capable of efficiently transporting electrons between the electrodes.

Even higher temperatures, above 1,900 degrees Fahrenheit, are required to operate a solid oxide fuel cell. In this technology, a chemical soup of exotic elements all claim a role. The electrolyte is a solid compound comprised of zirconium laced with a pinch of yttrium. This substance, unlike other electrolytes, is impervious to electrons and hydrogen ions but allows negatively charged oxygen ions to move freely between electrodes. The ion flow, therefore, is reversed compared with other fuel cells. Oxygen ions move from the cathode to the anode where they react with hydrogen to form water. The electrons released by this reaction flow through the external circuit as electricity back to the cathode, where they react with oxygen to form new ions. The cathode is usually made of lanthanum manganite laced with strontium and the anode is a composite of nickel and zirconia cermet. The entire fuel cell unit is encased by a composite ceramic container because conventional metal housings, steel and aluminum, are weakened at high temperatures.[33]

No contamination of the electrodes by the carbon compounds in air occurs in either solid oxide or molten carbonate fuel cells, so air can be used easily as the fuel oxidant. Moreover, in both cases, the water produced in the fuel cell appears as pressurized steam. This steam can be used to drive a turbine to generate electricity in addition to the electricity produced directly by the fuel cell. It is partly for this reason that fuel cell efficiencies greater than 60 percent are possible.

Another result of these technologies' high operating temperatures is the ability to use natural gas or methanol as a fuel rather than pure hydrogen. When using natural gas or methanol, the heat in the fuel cell breaks down the hydrocarbons into hydrogen and a variety of carbon compounds within the fuel cell. The fuel cell uses the hydrogen, and the carbon-containing by-products are removed. These by-products can be used as a supplemental, but not pollution-free, fuel for auxiliary power generation in a conventional engine. For example, the exhaust stream from molten carbonate or solid oxide fuel cells has a high enough temperature to raise steam in a heat exchanger. This steam can be run through a turbine to produce additional electric power.

Hydrogen Electric Hybrid Engines

Vehicles that combine on-board engines to generate power and electrical storage systems to store power are called electric hybrid vehicles. When a hydrogen-powered engine is used to generate the electricity, be it an internal combustion engine or a fuel cell, the vehicle is called a hydrogen electric hybrid vehicle. Interest in hybrid vehicle technology is growing because of the potential for such configurations to improve overall vehicle performance and help solve some of the technological problems facing vehicles powered completely by engines or completely by energy storage systems.[34]

Increasingly, development of hybrid vehicle technologies is moving to the forefront of advanced automotive research in the United States. As discussed in Chapter 8, electric hybrid programs are underway within the Departments of Energy and Defense, and hybrid technologies are included as part of President Clinton's Partnership for a New Generation of Vehicle, directed by the Department of Commerce. Most work on hybrids to date has focused on internal combustion engines powered by gasoline or alternative fuels other than hydrogen, for example, natural gas or methanol. Hydrogen electric hybrid engines, however, have been attracting increased attention since 1993.[35]

Benefits of Combining Engines and Electrical Storage Systems

Conventional vehicles are equipped with internal combustion engines that convert chemical energy in fuel on board a vehicle to another energy form that can be used for propulsion. Fuel cell engines use a different technology to accomplish the same purpose. Battery-powered electric vehicles, on the other hand, do not have an engine. They are equipped only with devices that temporarily store energy, originally generated outside the vehicle, until it is needed. Electricity is fed into the vehicle during refueling and later used to propel the vehicle.

By linking engines and electrical storage technologies, hybrid electric vehicles may prove to be lighter, smaller, more versatile, and better-performing than vehicles based solely on either engines or storage systems. For example, adding an onboard electrical generation capability to an electric battery-powered vehicle allows the size of the battery pack to be reduced. Because battery size and weight is one of the major problems facing battery-powered vehicles (see Chapter 5), the hybrid configuration, even including the weight of the engine, may be lighter than the all-battery mode using present battery technology.

Similarly, the presence of batteries in a hybrid vehicle permits a reduction in size of the engine, compared with an all-engine mode, because the energy in the batteries can be tapped to help propel the vehicle at high speeds. A typical 100-horsepower engine in a conventional automobile, for example, can be reduced to less than 50 horsepower because the energy stored in the batteries can provide the power needed during the relatively few moments when the vehicle is operating at high power demands. During normal driving patterns, a 100-horsepower engine in a conventional automobile is actually producing less than 30 horsepower about 80 percent of the time; even while cruising at 55 miles per hour on level terrain, the engine needs to produce less than 30 horsepower to keep the vehicle in motion.

In addition to reducing the size of the engine, which lessens engine weight and emissions, the engine in hydrogen electric hybrid vehicles can be engineered to operate at a steady speed, where the highest engine efficiencies can be achieved. When operating at a steady speed, for example, the efficiencies of internal combustion engines can increase above 30 percent.[36] Although this is less efficient than fuel cells, it is more than twice as efficient as the conventional engines used in non-hybrid automobiles today that must operate over a constantly changing power range in response to normal driving patterns. In the hybrid design, when the speed of the vehicle exceeds the power output of the steady-state engine, electricity can be drawn from the battery to provide the extra power that is needed. When the vehicle is idling or running at slow speeds, the excess power generated by the engine can be used to recharge the batteries.

Hybrids Using Hydrogen Combustion Engines or Fuel Cells

In hydrogen electric hybrids, hydrogen can be used either in an internal combustion engine or in a fuel cell. If used in a combustion engine, the energy released by burning hydrogen powers an electrical generator. A project to develop a hydrogen electric hybrid vehicle that uses an internal combustion engine to generate electricity is underway at Lawrence Livermore Laboratory in California. This vehicle is being designed to achieve a fuel efficiency of 96 miles per equivalent gallon of hydrogen relative to gasoline and to travel 300 miles between refueling, while emitting air pollutants within California's ultra-low emission standards. The engine technology is highly developed and could be available for widespread commercialization sooner than fuel-cell-powered vehicles.[37]

An alternative hydrogen electric hybrid technology combines a fuel cell engine with an electrical storage technology. The electricity either propels the vehicle immediately or is stored in batteries until needed. Several hydrogen electric hybrid systems are now under development that link a fuel cell and batteries. The phosphoric acid fuel cell bus sponsored by the Department of Energy (mentioned earlier) is the world's first operating hydrogen electric hybrid vehicle because it includes a large battery storage system capable of providing about half the power when the bus is operating at maximum speed. Most other fuel cell vehicles under development, including several of the buses being built in Europe as part of the Euro-Quebec Project discussed in Chapter 9, are also hybrids because they include battery systems to provide a significant portion of the vehicle's power.

Hydrogen-electric hybrids are not without problems. First, batteries may not be the ideal electrical storage medium. They may not respond fast enough to capture much of the energy from the vehicle's regenerative braking system. Flywheels with an alternator or ultracapacitors might be better long-term options. Second, undersized engines with storage may not be powerful enough to climb long, uphill grades. A conventional car is designed to maintain 65 miles per hour for a 3-mile-long grade of 7 percent. Hybrids designed thus far cannot achieve this level of performance.

Notes

1. James S. Cannon, *The Drive for Clean Air* (INFORM Inc., New York, 1989).
2. The primary source for this review of the early history of hydrogen vehicles was: Peter Hoffman, *The Forever Fuel: The Story of Hydrogen* (Westview Press, Boulder, CO, 1981), pp. 25-33.
3. Roger Billings, *The Hydrogen World View* (American Academy of Science, Independence, MO, 1991), pp. 78-102.
4. R. Billings and M. Sanchez, "Solid Polymer Fuel Cells: An Alternative to Batteries in Electric Vehicles – An Overview," *International Journal of Hydrogen Energy*, Vol. 20, No. 7, July 1995.

5. Hoffman, *The Forever Fuel..., op. cit.,* pp. 105-109.
6. Citations for the discussion of German and Japanese hydrogen programs appear in Chapter 9.
7. The principal sources of information for the section on hydrogen use in internal combustion engines include the following: Mark DeLuchi, "Hydrogen Vehicles: An Evaluation of Fuel Storage, Performance, Safety, Environmental Impacts and Cost," *International Journal of Hydrogen Energy*, Vol. 14, No. 2, February 1989; L. M. Das, "Combustion and Emission Characteristics of Hydrogen Engines," *Hydrogen Energy Progress X: Proceedings of the 10th World Hydrogen Energy Conference* (Cocoa Beach, FL, June 20-24, 1994); S. Lee *et al.,* "Combustion Characteristics of Intake Port Injection Type Hydrogen Fueled Engine," *International Journal of Hydrogen Energy*, Vol. 20, No. 4, April 1995; and R. Riley, *Alternative Cars in the 21st Century* (Society of Automotive Engineers, Warrendale, PA, 1994), pp. 171-194.
8. T. Petkow *et al.,* "An Outlook of Hydrogen as an Automotive Fuel," *International Journal of Hydrogen Energy*, Vol. 14, No. 7, July 1989.
9. R. Jorach and K. Prescher, "Development of a Low-NO_X Truck Hydrogen Engine with High Specific Power Output," *Hydrogen Energy Progress X: Proceedings of the 10th World Hydrogen Energy Conference* (Cocoa Beach, FL, June 20-24, 1994).
10. H. Mathur and P. Khajuria, "Performance and Emission Characteristics of Hydrogen Fueled Spark Ignition Engine," *International Journal of Hydrogen Energy*, Vol. 9, No. 8, August 1984.
11. Walter Stewart, "Hydrogen as a Vehicle Fuel," *Recent Developments in Hydrogen Vehicles* (CRC Press, Boca Raton, FL, 1986).
12. Y. Jamal and M. Wyszynski, "On-Board Generation of Hydrogen-Rich Gaseous Fuels – A Review," *International Journal of Hydrogen Energy*, Vol. 19, No. 2, July 1994.
13. *Ibid.*
14. V. Raman *et al.,* "Hythane – An Ultraclean Transportation Fuel;" and J. Wallace and A. Cattelan, "Hythane and CNG Fueled Engine Exhaust Emission Comparison," *Hydrogen Energy Progress X: Proceedings of the 10th World Hydrogen Energy Conference* (Cocoa Beach, FL, June 20-24, 1994). Hythane is a registered trademark of Hydrogen Consultants, Inc.
15. R. Hoekstra *et al.,* "Demonstration of Hydrogen Mixed Gas Vehicles," *Hydrogen Energy Progress X: Proceedings of the 10th World Hydrogen Energy Conference*" (Cocoa Beach, FL, June 20-24, 1994).

16. Wallace and Cattelan, "Hythane and CNG Fueled Engine Exhaust Emission Comparison," *op. cit.*

17. R. McAllister, "Sustainable Energy Future Will Begin with Uses of Hydrogen that Clean the Atmosphere," *Hydrogen Energy Progress X: Proceedings of the 10th World Hydrogen Energy Conference* (Cocoa Beach, FL, June 20-24, 1994); and interview with Robert Zweig, Clean Air Now, June 22, 1994.

18. DeLuchi, "Hydrogen Vehicles: An Evaluation of Fuel Storage, Performance, Safety, Environmental Impacts and Cost," *op. cit.*

19. Mark DeLuchi *et al., A Comparative Analysis of Future Transportation Fuels* (Institute of Transportation Studies, Berkeley, CA, 1987), p. 173.

20. *Ibid.*

21. "Debate Over Water Vapor Effect in Atmosphere," *The Hydrogen Letter*, April 1994.

22. "Principles of Fuel Cell Operation," and "Fundamentals of Fuel Cell," brochures from United States Department of Energy (Washington, DC, November 1992).

23. Principal sources of information for the section on fuel cell vehicle technology include the following: P. Skerrett, "Fuel Cell Update," *Popular Science*, June 1993; M. DeLuchi and J. Ogden, "Solar-Hydrogen Fuel-Cell Vehicles," *Transportation Research-A*, Vol. 27A, No. 3; "Fuel Cell Vehicles Will be Future Users of Alternative Fuels," *Clean Fuels Report*, April 1993; Dan McCosh, "Emerging Technologies for the Supercar," *Popular Science*, June 1994; D. Swan, "Fuel Cell Powered Electric Vehicles," SAE Technical Paper Series #189724, 1989; M. DeLuchi, *Hydrogen Fuel-Cell Vehicles* (Institute of Transportation Studies, Davis, CA, 1992); D. Simanaitis, "Technology Update: Fuel Cells," *Road & Track*, December 1994; and S. Kartha and P. Grimes, "Fuel Cells: Energy Conversion for the Next Century," *Physics Today*, November 1994.

24. *Compton's Interactive Encyclopedia 1992* (CD-ROM version); and Kartha and Grimes, "Fuel Cells: Energy Conversion for the Next Century," *op. cit.*

25. McCosh, "Emerging Technologies for the Supercar," *op. cit.*

26. "How Does a Fuel Cell Work," *Fuel Cell News*, Winter 1994.

27. D. Swan, "Fuel Cell Powered Electric Vehicles," *op. cit.*

28. "Tomorrow's Energy Solution was Born Over 150 Years Ago," brochure by Energy Partners, Inc. (West Palm Beach, FL, 1992).

29. C. Borroni-Bird, "Fuel Cells as a Future Automotive Powertrain," *Proceedings of the SAE Fuel Cells for Transportation TOPTEC Conference* (Santa Fe, NM, March 29, 1995).
30. "Texas A&M Research Cuts Fuel Cell Platinum Needs," *The Hydrogen Letter*, June 1994; and D. Oei, "PEM FC for Transportation: An Overview," *Proceedings of the SAE Fuel Cells for Transportation TOPTEC Conference* (Santa Fe, NM, March 29, 1995).
31. "Proton-Exchange Membrane (PEM) Fuels Cells," brochure by the United States Department of Energy (Washington, DC, November 1992); and N. Anand *et al.,* "Recent Progress in Proton Exchange Membrane Fuel Cells at Texas A&M University, *Hydrogen Energy Progress X: Proceedings of the 10th World Hydrogen Energy Conference* (Cocoa Beach, FL, June 20-24, 1994).
32. R. Green and V. Ahuja, "CO_2 Removal From Air For Alkaline Fuel Cells Operating with Liquid Hydrogen," *Hydrogen Energy Progress X: Proceedings of the 10th World Hydrogen Energy Conference* (Cocoa Beach, FL, June 20-24, 1994).
33. "Solid Oxide Fuel Cells," brochure by the United States Department of Energy (Washington, DC, November 1992).
34. "Volvo Asks Which is Better: The EV or the Hybrid?" *Clean Fuels Report*, February 1995.
35. Ray Smith, "The Hydrogen Hybrid Option," *Proceedings of the Workshop on Advanced Components for Electric and Hybrid Electric Vehicles* (Gaithersburg, MD, October 27-28, 1993); D. McCosh, "Emerging Technologies for the Supercar," *op. cit.*; and R. Riley, *Alternative Cars in the 21st Century, op. cit.,* pp. 201-266.
36. "Hydrogen Vehicles with Hydrogen Piston Engines Could Precede Fuel Cell Vehicles," *Clean Fuels Report,* November 1994.
37. "Livermore's Shock: Near-Term Hydrogen Use in Plausible," *The Hydrogen Letter*, April 1994.

Chapter 5

Storing Hydrogen in Vehicles

Of all the attributes that helped propel gasoline into its role as the predominant automotive fuel of the 20th century, none is more difficult for alternative fuels to match than the ease with which gasoline can be carried on board a vehicle. As a liquid, gasoline can be stored conveniently in a thin-shelled, lightweight tank tucked out of sight underneath a car. Refueling is a simple matter of pumping more liquid into the tank, which, on most cars, can hold enough fuel to travel more than 300 miles without refilling. Continued reliance on gasoline may be increasingly untenable with regard to air pollution, fuel supply, safety, and energy efficiency, but with regard to fuel storage characteristics, gasoline retains a great advantage.

The development status of hydrogen fuel storage systems suitable for use on board vehicles is primitive, but there is progress. Hydrogen enjoys an advantage over gasoline that may provide the key to fuel storage. The light weight of the fuel itself and the high efficiency of hydrogen fuel cell engines mean that the amount and weight of fuel needed on board a hydrogen vehicle is only a fraction of the amount of gasoline stored in a conventional car.

The principal barrier to technological advancement is not some impossible technical obstacle; the main barrier, rather, is the lack of adequate public- and private-sector financial support. Within a decade, substantial research investment would likely lead to development of a variety of vehicles fueled by hydrogen and performing as well or better than the gasoline vehicles of today.

Three Critical Issues: Weight, Space, and Refueling Speed

The viability of any fuel storage system is affected by the weight of the storage system, the volume it requires, and the ease with which it can be refueled. In order to be commercially viable, vehicles powered by any of the alternatives to gasoline, including hydrogen, must have fuel storage systems to provide an adequate driving range without adding significant weight or space requirements. The three most developed options for storing hydrogen entail compressing the gas, converting it to a liquid, and binding it to solids via metal hydride or carbon adsorbent technologies. Other options involve storing hydrogen in the molecular structure of other chemicals; the hydrogen is then released when chemical reactions break down these chemicals. Developing the technology to deliver fuel to and from the system as quickly and easily as a gasoline vehicle presents a second, related challenge.

Weight of Fuel Storage Systems

Three factors influence the total weight of an energy storage system onboard a vehicle:

- Weight and energy content of the energy source itself.
- Weight of the hardware that contains the energy source.
- Propulsion efficiency of the system.

In terms of weight, hydrogen surpasses two of the three other leading energy sources vying for the automotive fuel market – gasoline and natural gas – as shown in Table 1. It takes only 16 pounds of hydrogen to supply 1 million Btu of energy, compared with 49 and 45 pounds of gasoline and natural gas, respectively. However, the fourth energy source, electricity, is the lightest of all. Electricity is obtained from energy held in electrons that weigh almost nothing compared with the atoms that hold energy in chemical bonds in the other fuels. Thus electrical energy itself weighs virtually nothing, although the electrical storage systems can be very heavy, as explained below.

Table 1. Weight of Various Automotive Fuels

Fuel	Energy Content (pounds per million Btu)	Weight of 15 Gallons or Gallon Equivalents* (pounds)	Weight Relative to Gasoline
Gasoline	49	83	–
Natural Gas	45	76	-8%
Hydrogen	16	27	-67%
Electricity	0	0	-100%

*A gallon equivalent is an amount of fuel containing the energy of 1 gallon of gasoline (114,000 Btu).

Source: INFORM.

Yet when the weight of the entire fuel storage system is taken into account, the respective merits of the four fuels dramatically reverse. As shown in Table 2, the best lead acid battery system on the market is 180 times heavier than a gasoline storage system. The storage system for gasoline, including 15 gallons of fuel and the standard thin-walled sheet metal or plastic tank that holds it, weighs only about 150 pounds. A battery system holding as much energy as a 15-gallon tank of gasoline would weigh 27,000 pounds, about 10 times the entire weight of an average car.

Because of the heavy weight of conventional batteries, the first generation of modern electric vehicle prototypes – for example, the General Motors Impact – generally carry on board less than one equivalent gallon of fuel in the form of electricity.[1] By carrying only a small amount of fuel, the battery weight can be held to a manageable level; for example, about 870 pounds in the case of the Impact.[2] Although this electrical energy can be used much more efficiently in electric motors compared with conventional engines, nonetheless the result is greatly reduced driving range. Most prototype electric cars on the road today have a driving range of less than 100 miles. Similarly, the state-of-the-art electric bus unveiled by the Chattanooga Area Regional Transportation Authority in Washington, DC, on Earth Day 1994 is equipped with a pack of batteries weighing 2.5 tons – more than three times heavier than the vehicle's old diesel engine – but the bus has a maximum driving range of only 75 miles before recharging is necessary.[3]

Table 2. Weight of Fuel Storage Systems Containing Equal Energy

Fuel	Fuel Storage System Hardware (pounds per gallon or gallon equivalent*)	Fuel Storage System Hardware Containing 15 Gallons or Gallon Equivalents (pounds)	Weight Relative to Gasoline
Gasoline	10	150	–
Compressed Natural Gas	20	300	2x
Compressed Hydrogen	60	900	6x
Liquefied Hydrogen	30	450	3x
Hydrogen Hydrides	120	1,800	12x
Lead Acid Battery	1,800	27,000	180x
Battery Mid-term	900	13,500	90x
Advanced Battery	180	2,700	18x

Source: INFORM.

The Advanced Battery Consortium (ABC), a partnership between the United States Department of Energy and the three major domestic auto manufacturers, initiated a four-year, $260 million research program in early 1991 to address some of the shortcomings of battery technology, including their heavy weight.[4] The ABC has established a mid-term goal of increasing the energy-carrying capability of a battery to the point where it is only 90 times as heavy as a gasoline system. The ABC's long-term goal is a battery 18 times heavier than a gasoline system, still much heavier than currently available hydrogen storage systems. Several prototype batteries already on the market are capable of meeting the mid-term goal, but a commercially viable battery capable of meeting the long-term goal remains elusive.

Options for storing natural gas and hydrogen

The storage systems for the gaseous fuels hydrogen and natural gas are significantly heavier than a gasoline storage system containing the same amount of energy. But the hydrogen and the natural gas storage systems are not nearly as heavy as batteries. The most common method of storing natural gas on board a vehicle is in compressed gas storage cylinders. Most new natural gas vehicles today use metal or plastic cylinders wrapped with high-strength fiberglass or carbon fibers. To save space, the fuel is highly compressed, usually to 3,000 pounds per square inch. To withstand such pressure, the cylinder walls must be much thicker and heavier than the walls of a gasoline tank. Even the most lightweight natural gas tanks on the market today result in a fuel system weight that is at least twice as heavy as a gasoline system: 300 pounds for a system containing the energy equivalent to 15 gallons of gasoline. Most vehicles on the road today are equipped with less costly storage cylinders weighing about three times more than a comparable gasoline system.

The weight disadvantage of hydrogen systems is much greater than that of natural gas. A compressed gas fuel storage system for hydrogen is at least six times heavier than a gasoline system. At 900 pounds, a compressed hydrogen system may stretch the limits for viability in automotive applications unless vehicles with lesser fuel demand – such as higher-efficiency fuel cell vehicles – can be used. Development of new ultra-light materials that can withstand higher pressures, up to 5,000 pounds per square inch, could reduce this weight penalty, but this technology currently is not commercially available.

However, the fuel storage system weight for hydrogen can be significantly reduced if the hydrogen is liquefied (instead of compressed) and stored aboard the vehicle in specially insulated containers. Liquefying hydrogen can reduce the fuel system weight to half that of a compressed-hydrogen system. A liquefied-hydrogen system is only three times as heavy as a gasoline system holding the same amount of energy. However, as explained later in this chapter, hydrogen liquefies only at extremely low, or cryogenic, temperatures – less than 400 degrees Fahrenheit below zero. The process is expensive and presents engineering and safety problems related to handling super-cold fluids.

A third storage option for hydrogen involves reversibly reacting hydrogen with metal alloys. The reaction bonds the hydrogen to the metal, producing what are called metal hydrides, and releases heat in the process. When fuel is needed, the hydrides are heated, which in turn releases the hydrogen. Metal hydride storage systems avoid the high-pressure requirement of compressed gas systems and the difficulty of producing and handling liquefied hydrogen. However, the hydride metal itself is heavy: Such a storage system is about 12 times as heavy as a gasoline system holding the same amount of energy.

Several other hydrogen energy storage options in addition to the three main technologies mentioned above are described later in this chapter. They, too, are heavier than gasoline storage systems.

Options for storing electricity

Two electrical storage technologies that are potential alternatives or supplements to electric batteries are flywheels and ultracapacitors.[5] Flywheels store electricity mechanically rather than chemically. They consist of rapidly spinning wheels encased in a nearly friction-free, vacuum-sealed container. The energy of rotation – spin rates as high as 200,000 revolutions are possible – can be tapped to generate electricity as needed. As the energy is depleted, the speed of the spinning wheels decreases; refueling involves using electricity to boost the speed of rotation. The weight of flywheel systems is half the weight of lead acid batteries, but it is still about 90 times as heavy as a gasoline system.

Ultracapacitors offer a second electrical storage option; they store electricity on a conducting plate separated from a second plate by a vacuum. Nearly every electronic device, including televisions and radios, uses capacitors, so the basic technology is advanced compared with that of flywheels. However, scaling up capacitors to store much larger amounts of electricity entails a number of technical and engineering difficulties. Ultracapacitors excel at delivering large, short bursts of energy, but their ability to deliver sustained energy flows for long periods is limited. As a result, current ultracapacitor technology is much heavier than any energy storage technology now under study for automotive applications; for example, they are 5 to 10 times as heavy as lead acid batteries.

Propulsion efficiency

The ultimate weight of a fuel storage system onboard a vehicle is affected by more than the weight of the fuel, the weight of the hardware, and the energy content of the fuel. The main issue of concern to a motorist is how far the vehicle can be driven between stops to refuel, which is a function not only of the energy in the fuel, but of how efficiently the system delivers that energy to propel the vehicle. Electric propulsion systems are much more efficient than internal combustion engine-powered vehicles. A vehicle that is propelled by electricity produced in a hydrogen fuel cell is approximately three times as energy efficient as a vehicle powered by an internal combustion engine; that is, three times more of the energy in the fuel reaches the wheels of the vehicle. An electric vehicle fueled by electricity stored in batteries is about four times as efficient as an internal combustion engine. Thus, the amount of fuel, hence the weight of the fuel storage system, can be slashed by two-thirds in the case of a hydrogen fuel cell vehicle and by three-quarters in the case of a battery-powered vehicle while still providing enough energy to travel as far as a vehicle fueled with 15 gallons of gasoline.[6]

Table 3 shows the weight of various fuel storage systems needed to provide comparable driving ranges for vehicles. When used in internal combustion engines, the weight disadvantages of natural gas and hydrogen relative to gasoline remain basically the same as in Table 2, because they are all being used in the same type of propulsion system. When used in a fuel cell, however, the weight disadvantage for hydrogen is dramatically reduced compared with its use in an internal combustion engine. The weight of a liquefied hydrogen system becomes about the same as the weight of a gasoline system offering a similar driving range. The compressed hydrogen system becomes about twice as heavy as the gasoline system, offering about the same disadvantage as state-of-the-art natural gas vehicles today. Even a hydride system, the heaviest hydrogen storage option, is reduced in weight to only four times that of a gasoline system when the hydrogen is used in a fuel cell.

By factoring in propulsion efficiency, the weight penalty of batteries is also drastically reduced. Unlike the hydrogen fuel cell case, however, the weight for battery systems is still generally much heavier than that for

Table 3. Weight of Fuel Storage Systems for Driving Range Equal to Gasoline

	Weight of Fuel Storage System (pounds)	**Relative to Gasoline**
Internal Combustion		
Gasoline	150	–
Compressed Natural Gas	300	2x
Compressed Hydrogen	900	6x
Liquefied Hydrogen	450	3x
Hydrogen Hydrides	1,800	12x
Fuel Cell		
Compressed Hydrogen	300	2x
Liquefied Hydrogen	150	same
Hydrogen Hydrides	600	4x
Battery		
Lead Acid Battery	6,750	45x
Battery Mid-term	3,375	22.5x
Advanced Battery	675	4.5x

** A typical gasoline tank holds 15 gallons of fuel and offers a driving range of 300 miles.*

Source: INFORM.

gasoline systems. A lead acid system, for example, is 45 times heavier than a gasoline system offering the same driving range. This penalty can be further cut in half, to 22.5 times heavier than gasoline, by using batteries that meet the mid-term objectives of the Advanced Battery Consortium. But even if the long-term goals of the ABC are met, batteries would still be 4.5 times as heavy as the weight of the heaviest hydrogen storage system in use today.

Heavy weight remains a problem facing all alternative fuels. The weight penalty for battery-powered electric vehicles is onerous, the high efficiencies of electric systems notwithstanding. The weight disadvantage for hydrogen is potentially crippling when proposed for use in internal combustion engines. However, when hydrogen is used in efficient fuel cell electric vehicles, the weight disadvantage for hydrogen becomes comparable to the challenge now being met successfully by natural gas vehicles.

Table 4. Volume of Fuel Storage Systems for Driving Range Equal to a 15-Gallon Gasoline Tank

	Fuel System Volume (cubic feet)	Relative to Gasoline
Internal Combustion		
Gasoline	2.67	–
Compressed Natural Gas	10.69	4x
Compressed Hydrogen	32.08	12x
Liquefied Hydrogen	10.69	4x
Hydrogen Hydrides	24.03	9X
Fuel Cell		
Compressed Hydrogen	10.69	4x
Liquefied Hydrogen	3.34	1.2x
Hydrogen Hydrides	8.02	3x
Battery		
Lead Acid Battery	45.45	17x
Battery Mid-term	32.08	12x
Advanced Battery	13.37	5x

Source: INFORM.

Volume of Fuel Storage

In general, a heavier fuel storage system also means a bulkier system. However, larger space requirements are not quite as troublesome an issue as weight. Additional weight slows down a vehicle, reduces fuel efficiency, impairs braking, and impedes vehicle handling. By contrast, vehicles can be redesigned to provide considerably more fuel storage space without adversely affecting vehicle performance.

As shown in Table 4, in terms of storage space, hydrogen can best compete with gasoline-powered vehicles when it is used in fuel cells to power electric vehicles. A liquefied hydrogen fuel cell storage system takes up only slightly more space than a gasoline system offering the same driving range. Compressed hydrogen and hydride storage systems are only three and four times bulkier, respectively.

Hydrogen fuel cell storage systems also compete well with electric batteries, taking up less space than the most advanced projections of the battery industry. Even the best lead acid batteries available today are 17 times larger than a comparable gasoline system. Batteries of the future may be able to reduce that volume penalty to 12 times larger if the mid-term goals of the Advanced Battery Consortium are met, but they will still be 5 times larger even if the long-term goals for advanced battery technology are achieved.

Both natural gas and hydrogen face volume disadvantages compared with gasoline systems when used in internal combustion engines, as shown in Table 4. A compressed natural gas storage system is about four times as bulky as a gasoline system containing an equal amount of energy. Hydrogen destined for use in an internal combustion engine also suffers a disadvantage, requiring between four and twelve times the volume of a gasoline system, depending on how it is stored. As in the case of weight, hydrogen's disadvantage vis-a-vis gasoline is minimized when the hydrogen is stored in a liquefied state. Storing hydrogen in the form of hydrides also lowers the space requirements with respect to a compressed hydrogen system.

Refueling Speed and Cost

Gasoline refueling is so easy and quick that it often takes longer to stand in line and pay than the few minutes it takes to refuel the vehicle. Refueling with alternative transportation fuels today is generally more complex, more costly, and takes longer than with gasoline. Each of the alternative fuels under discussion faces a major disadvantage compared with gasoline: compressed fuel systems for natural gas and hydrogen require expensive equipment to achieve high pressure levels, and liquefied hydrogen systems require expensive liquefaction equipment and exacting procedures to handle super-cold fluids (possibly necessitating the return of the service station attendant). Hydrogen hydrides and electric batteries are encumbered by the slow rate of chemical reactions during the refueling or recharging process.

Internal combustion engine-powered vehicles with compressed natural gas and fuel cell-powered vehicles with compressed hydrogen come closest to gasoline in terms of refueling speed. The best "fast-fill" natural gas refueling stations equipped with powerful compressors now come close to matching gasoline's refueling time, although they still take slightly longer, typically 5 minutes in total. The same natural gas compressor technology can be used to refuel a hydrogen vehicle powered by an internal combustion engine, but refueling with compressed hydrogen takes about three times as long as refueling with natural gas, owing to the need to compress a greater volume of hydrogen to match the energy content of natural gas. If the hydrogen vehicle is powered by a more efficient fuel cell, less fuel is required and the refueling time for hydrogen is approximately the same as that needed for a natural gas vehicle.

Refueling with liquefied hydrogen entails certain disadvantages, taking at least five times as long as refueling with gasoline. Although both fuels are liquids, the extreme cold temperature of liquefied hydrogen requires special handling procedures that delay the refueling process. Also, liquefied hydrogen is not as energy-dense as gasoline, so more liquid must be transferred to the vehicle to provide the energy equivalent to a tank of gasoline.[7]

Refueling a hydrogen hydride storage system is similarly constrained, although in the case of hydride systems it is the heat generated as hydrogen reacts with the hydride, rather than the cold, that causes the delay. The faster the heat is released from the system, the more quickly hydrogen can be absorbed by the metal. Like liquefied hydrogen refueling, a hydride refueling system takes at least five times as long as a gasoline system.

Battery refueling: Quick recharging a problem

As with weight and volume considerations, battery technology currently suffers compared with hydrogen fuel storage technologies with regard to speed of refueling. Batteries store energy in the form of chemicals: When the chemicals are used up, the battery must be recharged with electricity from an outside source to reverse the chemical reactions and restore the reactants. Recharging battery-equipped electric vehicles is limited by the rate at which the chemical reactions occur; with current technology, recharging usually takes 6 to 10 hours. "Quick-charge" battery recharging systems are being developed, but they operate at very high electrical currents, which raise safety concerns and make them more costly to operate than overnight charging systems.

Even advanced recharging systems are slow compared with gasoline refueling. In February 1994, for example, *The New York Times* reported a "breakthrough technique" capable of speeding the rate of battery recharge on electric vehicles; in just under 19 minutes, the new technique "crammed" electricity equivalent in energy content to half a gallon of gasoline into a battery pack.[8]

Another problem with batteries is that the number of times they can be recharged is limited because the chemicals they contain gradually decompose as the batteries undergo cycles of charging and discharging. The current generation of lead acid batteries needs to be replaced after the vehicles travel less than 20,000 miles, or about two years of use for an average vehicle. The mid-term goal of the Advanced Battery Consortium is to develop a battery with a five-year life; the long-term goal is to produce a battery that can last the lifetime of an average car without replacement.[9]

On the other hand, recharging batteries with electricity and refueling with natural gas offer one potential advantage today over gasoline in that these processes can take place while a vehicle sits idle at a home or business – albeit slowly. "Slow-charging" electric batteries and "slow-filling" natural gas vehicles can be accomplished overnight or during business hours, whenever the vehicles are not in use.

Provided that a home or business is served by electricity, and if time is not of the essence, electric battery vehicles can be plugged into a modified wall socket and refueled conveniently over a period of hours. Overnight refueling of electric batteries can take advantage of the lowest "off-peak" rates available for electricity. Surveys of driving patterns show that most vehicles are driven fewer miles per day than can be provided by batteries now installed in electric vehicles. This means that in many if not most "real-world" driving conditions, there is no need for midday, fast refueling. The disadvantage of home or business refueling, of course, is that the vehicle cannot be used during the long periods while refueling takes place. Such circumstances may be rare for many people, but they represent a limitation on vehicle availability that is not present with conventional vehicles.

Hydrogen Vehicle Fuel Storage Technologies

Two of the most common methods for storing natural gas in vehicles, compression and liquefaction, can be easily adapted for storing hydrogen. A third option entails providing an onboard system that would release hydrogen from another chemical substance, such as methanol.[10] Other technologies that use metals or activated carbons to hold hydrogen take advantage of hydrogen's propensity to bond with a variety of solid materials. Another technology takes advantage of the reaction between water and iron, known as rusting, which releases hydrogen as a by-product.

These five technologies are discussed in the order of their current development and potential for application to hydrogen vehicles:

- Compressed hydrogen
- Liquid hydrogen
- Binding hydrogen to solids
- Rusting sponge iron to release hydrogen
- Chemical hydrogen carriers.

Adapting Compressed Natural Gas Technology for Hydrogen

About 99 percent of the 750,000 natural gas vehicles on the road worldwide use compressed gas storage cylinders to hold the fuel supply. Although this technology is directly applicable to hydrogen, its heavy weight and large volume requirements, which are difficult but manageable for natural gas, may become unwieldy for hydrogen when used in an internal combustion engine. However, the greater efficiency of fuel cell engines may allow vehicles to use compressed hydrogen storage systems with a weight and volume comparable to a natural gas system.

In the United States there are about 50,000 natural gas vehicles, nearly all using compressed fuel. To service these vehicles, more than 1,000 compressed natural gas refueling stations have been built, with two more opening each week. This compares with approximately 200,000 gasoline refueling stations nationwide, which service more than 190 million gasoline and diesel-powered vehicles.[11]

Fast and slow refueling

At one of the nation's fast-fill natural gas refueling stations, a car can normally be refueled in 5 minutes, only slightly longer than the time it takes to refuel a conventional automobile with gasoline. At a fast-fill station, natural gas is typically compressed to about 3,600 pounds per square inch (psi) using heavy-duty industrial compressors. The highly compressed gas is stored in large cylinders until it is needed. Refueling entails connecting a hose from one of these storage cylinders to a refueling port on the vehicle that, in turn, leads to onboard fuel storage cylinders. When connected, the natural gas flows from the high-pressure storage tanks into the vehicle's cylinders. When pressure in the on-board cylinder reaches 3,000 psi, the system shuts off and refueling is completed. The compressors replenish the stored natural gas in the 3,600-psi tank at the refueling station.

In slow-fill natural gas refueling, smaller compressors gradually pump fuel from a distribution pipeline, which typically delivers gas at 5 to 15 psi, into a vehicle's storage cylinders. Slow-fill refueling systems generally compress natural gas at a rate of 1 to 3 gallons per hour, so refueling can take several hours, or overnight. As the compressor works, the gas pressure in the tank slowly increases until 3,000 psi is reached and the system shuts off. An advantage of slow-fill refueling is that small compressors can do the job, thereby cutting roughly in half the capital costs of the large compressors needed for fast refueling and reducing the operating costs.

Moreover, home slow-fill systems are more convenient than driving to gasoline stations to refuel. Extensive natural gas home refueling could reduce the need for many of the refueling stations that now mar the roadside landscape. If a home or business is served by natural gas, which includes about 25 percent of all buildings in the United States, natural gas vehicles can be connected to slow-fill compressors linked to gas distribution pipelines that extend into the buildings.

Three types of storage cylinders

There are three types of natural gas storage cylinders currently in use:

- Pressed steel
- Wrapped aluminum
- All-composite.

Cylinders made of pressed steel were the industry standard until the mid-1980s. They are still used widely to store a variety of compressed gases at stationary locations, but their heavy weight limits their usefulness in automobiles. A steel cylinder holding an amount of natural gas with an energy content equal to 5 gallons of gasoline, called "equivalent gallons," weighs about 150 pounds.

In the late 1980s, a second type of fuel storage cylinder entered the market. It is based on lighter-weight aluminum wrapped with strings of very high-strength carbon or fiberglass. The wrapping adds strength to the cylinder, thereby permitting a reduction in the thickness of the aluminum sheet. The result is a cylinder weighing about half as much as a pressed steel cylinder, or about 75 pounds for a size holding the equivalent of 5 gallons of gasoline.

The third, and newest, cylinder type uses lightweight, very strong ceramics for the shell, which is then wrapped with carbon or fiberglass strings. Called all-composite cylinders, they reduce weight to about 50 pounds per 5 equivalent gallon cylinder, 33 percent lighter than with wrapped aluminum cylinders and 70 percent lighter than with pressed steel. All-composites first entered the commercial market for natural gas vehicle cylinders in 1993 and are gaining quickly in popularity.

Even lightweight natural gas storage cylinders are heavy compared with the thin-walled gasoline tanks in common use. A 15-gallon gasoline tank filled with fuel weighs about 150 pounds, more than half of which is the fuel itself; a comparable compressed natural gas storage system using all-composite cylinders weighs 300 pounds, including tank brackets and the natural gas, while a wrapped-aluminum cylinder system weighs 450 pounds. Because of high cylinder weight, the search for ever-lighter materials for natural gas storage cylinders continues. In June 1994, for example, a partnership between the United States Department of Energy and the three major auto manufacturers was formed to develop lighter and cheaper natural gas storage cylinders. Called the Natural Gas Vehicles Technology Partnership, the group hopes to reduce the cost of producing lightweight tanks by 50 percent within four years.[12]

High volume and pressure requirements

The problem of the weight of storage cylinders is made more difficult by the low volumetric density of natural gas. As discussed above, even when compressed to 3,000 psi, natural gas occupies four times as much space as a comparable amount of gasoline. One way to lessen the space requirements of natural gas is to increase the amount of fuel in the cylinders by raising the pressure. A decade ago, fuel cylinders were routinely filled to only 2,400 psi. The industry norm today is 3,000 psi, increasing the amount of fuel stored in a cylinder by 20 percent, and some cylinders are now approved for use at 3,600 psi.[13]

Increasing the pressure also increases the refueling cost, however. Compression to 3,000 psi now adds up to 20¢ to the cost of an equivalent gallon of natural gas, half for the capital cost of the compressor and half for the energy to run the system. The higher the storage pressure, the more rapidly compression costs increase. Raising the industry standard for natural gas vehicle systems to 3,600 psi could add another 3¢ per equivalent gallon to the compression bill.[14]

An alternative to increasing the pressure is to redesign gasoline vehicles so that more space is available for fuel storage without compromising passenger or cargo storage. This may be the long-term answer to the volume issue facing the natural gas vehicle industry; some vehicles are already being designed from scratch to accommodate fuel storage cylinders. For example, buses with nine large compressed gas storage cylinders mounted unobtrusively on the roof of urban buses entered commercial production by Bus Industries of America, Inc., in 1991.

Hydrogen applications of compression technology

Because hydrogen is only one-third as energy-dense as natural gas (per unit of volume), a storage capacity three times as large is needed to hold an energy-equivalent amount of hydrogen. Unless the storage pressure is increased or the demand for fuel by the engine decreased, this means that a compressed hydrogen storage system would weigh nearly half a ton and occupy more than 9 times as much space as a gasoline tank.

Nevertheless, there is a place for compression gas technology in the development of hydrogen vehicles today and potentially in the future. Several demonstration vehicles now operating in the United States and abroad, discussed in Chapters 8 and 9, use compressed hydrogen storage systems because the technology is available readily through the growing market for natural gas vehicles. Equipping a vehicle with compressed gas storage cylinders holding a couple of equivalent gallons of hydrogen occupies only a little more space than a conventional gasoline tank and weighs about the same. This amount of fuel is sufficient for many vehicle testing and demonstration purposes.

Compression technology becomes more attractive when used in a fuel cell-powered vehicle. By tripling the fuel efficiency of the engine, as shown in Tables 3 and 4, fuel cells brings the space and weight requirements of a hydrogen storage system down to the levels of today's natural gas vehicles.

In today's vehicle markets, it is impossible to ignore the demand for vehicles with greater driving ranges between refueling stops. One option for increasing onboard fuel capacity while retaining use of compression storage technology is to increase the operating pressures of compressed hydrogen cylinders. Spherical aluminum-wrapped and all-composite storage systems holding hydrogen compressed to 10,000 psi were developed in the 1960s and have been used in the space program ever since. By tripling the pressure used in standard natural gas storage cylinders, the space requirements of compressed hydrogen systems become about equal with the natural gas storage cylinders in use today. However, the compression costs and the hydrogen storage spheres themselves would be much more expensive than conventional cylinders.

Finally, more exotic fuel storage technologies for compressed hydrogen, which have been demonstrated under laboratory conditions, may find an application in hydrogen vehicles over the long term. One such system, called microballoon or microcapsule storage, uses tiny, hollow balls of very strong glass to capture and release hydrogen at high pressure.[15] Glass microcapsules, each smaller than a grain of sand, can be produced in specially designed furnaces. These capsules are very strong, capable of holding hydrogen at pressures exceeding 15,000 pounds psi. At room temperature they are impermeable to hydrogen, but the gas freely passes through their walls at temperatures about 2,500 degrees Fahrenheit. If compressed hydrogen is introduced into an oven containing hot glass microcapsules under high pressure, the hydrogen will fill each capsule. If the capsules are cooled without reducing the pressure, hydrogen will become trapped inside the capsules as they return to their normal state of impermeability.

Once filled with hydrogen and cooled, the pressure can be reduced and the capsules can then be stored in a lightweight tank similar to a gasoline tank on board a vehicle. To release the hydrogen to power the vehicle, the capsules in the fuel tank must be gradually heated until hydrogen can once again pass through their walls. The empty capsules can be removed and refilled with hydrogen. There are a number of technical problems with microcapsule systems, including the tendency of the glass to break if roughly handled, and these problems must be resolved before they can be applied to transportation applications. No vehicles are currently operating with a microcapsule storage system.

Transferring Technology between Liquefied Natural Gas and Hydrogen

When natural gas and hydrogen are in a liquid state, they are much more energy-dense per unit of volume than when they are gases, allowing significant reductions in storage space over compressed gas systems. However, the extremely cold temperatures needed to maintain these substances in liquid form present a costly technical challenge for vehicle use. Nonetheless, liquid hydrogen stored for use in a fuel cell engine offers the possibility of an alternative fuel storage system with weight and space requirements similar to that of a gasoline system.

As a result of decades of production for the space program, the technology to liquefy hydrogen is well established and reliable. It is based on the scientific principle that when compressed gases are allowed to quickly expand, their temperature drops. Blowing on your hand demonstrates this phenomenon. The air leaves your lungs at about body temperature, 98 degrees Fahrenheit, but it feels cool on your hand as a result of the expansion that takes place after it leaves your mouth. Conventional liquefaction processes involve a series of compression and expansion steps, each one lowering the temperature of the gas until liquefaction occurs.

Learning from liquefied natural gas use

In spite of the complexities of attaining the extremely cold temperature required to liquefy natural gas, -258 degrees Fahrenheit, there has been a huge upswing in liquefied natural gas vehicle projects in the United States since 1992. More than 30 liquefied natural gas vehicle projects are underway in the United States, mostly undertaken by operators of trucks and buses. Several bus companies in Texas have made a commitment to operate hundreds of buses powered by liquefied natural gas, and several other transit operators are converting lesser numbers of buses. In addition, operators of long-haul trucks, including Roadway Express, are testing liquefied natural gas, which offers a higher energy density compared with the fuel in other forms. Despite these programs, less than 1 percent of the natural gas vehicles on the road worldwide today are fueled by liquefied natural gas. As operators and technicians gain more practical experience in the use of liquefied natural gas in automotive applications, this knowledge will likely help them further the development of the technology to use liquefied hydrogen.[16]

Applications of liquid hydrogen

The same basic cryogenic technology used to liquefy natural gas is also used to liquefy hydrogen, although much colder temperatures must be achieved before hydrogen gas condenses into a liquid. Liquefied hydrogen gas is one of the coldest substances ever produced by human beings, requiring a temperature of -423 degrees Fahrenheit; this is only about 36 degrees Fahrenheit above absolute zero, an unattainable temperature at which all molecular motion stops. The only element on earth with a lower liquefaction temperature, more commonly known as a boiling point, is helium. Despite the difficulties in attaining and maintaining such frigid temperatures, liquefied hydrogen has been produced in quantities sufficient to provide the principal fuel for the space program, and it has already been successfully used as an automotive fuel in a number of demonstration vehicles.[17]

In the United States, scientists at the Los Alamos National Laboratory in New Mexico built a hydrogen vehicle powered by liquefied hydrogen in 1979 and operated it for 17 months, until the project's funding ended. In

Germany, most of the prototype hydrogen vehicles built by BMW, Inc. during the 1980s and early 1990s were or are fueled by liquefied hydrogen. In Japan, most of the hydrogen vehicles developed at the Musashi Institute of Technology are fueled by liquefied hydrogen, including the currently operating Musashi-8 sports car and the Musashi-9 refrigerated food delivery truck – which is cooled by the liquefied fuel.

Advantages of liquid fuels

Because liquefied natural gas and hydrogen enjoy the benefit of enormously increased energy density compared with their gaseous forms, they consume less than half the space required by compressed gas or hydride storage systems – although they are still bulkier than gasoline. For example, a liquefied hydrogen storage system takes up only 4 times more space than a gasoline system containing the same energy content.

Another advantage is that liquids can be stored easily at low pressure, thereby reducing the need for costly compressor operations and for very strong and consequently thick and heavy storage containers. The insulated tanks, nevertheless, are still heavy compared with gasoline tanks. A liquefied hydrogen storage system weighs about three times as much as a gasoline system containing comparable energy. If the hydrogen were destined to be used in a fuel cell engine with a fuel efficiency three times that of an internal combustion engine, liquefied hydrogen storage would be comparable to or only marginally larger than a gasoline system with regard to weight and volume.

Handling super-cold fluids

Once liquefied, both natural gas and hydrogen are stored in similar, specially designed, super-insulated tanks to slow the gradual warming of very cold substances that causes a gradual evaporation of the liquid, called "boil-off." Handling both fuels in a liquefied state requires special procedures because the extreme cold freezes normal mechanical equipment, such as refueling station pumps, rendering them inoperable. Moreover, the temporary removal of the fuels from the insulated storage tanks during refueling rapidly increases the boil-off rate and creates safety problems, until the fuel is deposited in the vehicle's tank. A drop of either fuel in a liquefied

Super-insulated tanks hold liquefied hydrogen fuel at extremely low temperatures. Photo: BMW

state will cause a severe burn if it lands on bare skin, for example, raising concern for refueling station personnel handling super-cold fluids.

High production cost and energy use

Conventional liquefaction is very expensive and energy-consuming. Up to 40 percent of the energy in hydrogen is used during the liquefaction process.[18] The process adds about $7 per million Btu to the cost of the hydrogen that is liquefied, an amount almost equivalent to the price of gasoline selling at $1 per gallon. However, a new technology in the early stages of development, called magnetic refrigeration, holds the promise to cut both energy loss and the cost of conventional liquefaction in half. Although demonstrated in the laboratory, it is not yet in commercial application.[19]

Storage challenges

Once liquefied, hydrogen must be kept cold until it is used. Double-walled, heavily insulated tanks called Dewars are used for this purpose. Dewars maintain a vacuum in the inner space where hydrogen is stored because most of the molecules in air would solidify at the temperature of liquefied hydrogen. To improve insulation, the tanks also maintain a vacuum in the space between the inner and outer walls. This space is also filled with super-insulating materials, such as alternating layers of aluminized mylar and silk. Spherical tanks are often used to hold liquefied hydrogen because this shape minimizes the area of contact between the walls and the fuel where heat exchange can take place. Moreover, the larger the tank, the lower the ratio between wall surface and the volume of fuel.

As liquefied hydrogen sits in storage near the site of end use, whether on board a vehicle or at a refueling station, a gradual warming is inevitable. As the hydrogen slowly boils off, the pressure of hydrogen gas in the tank increases. Unless this gas is drawn off periodically, the pressure building up would eventually burst the tank; as a safety precaution, hydrogen Dewars are usually equipped with a burst valve that automatically releases fuel through a vent whenever the pressure in the tank rises above 50 pounds per square inch.

As long as the hydrogen fuel is frequently used, boil-off is not a problem because accumulated gases are drawn off first when fuel is removed. Dewars commonly in use are so well insulated that fuel can sit unused for 5 to 7 days before the internal pressure reaches problematic levels. This translates to a boil-off rate of less than 2 percent of the fuel per day. Dewars developed as part of the space program can reduce boil-off to less than 0.5 percent per day. Such technologies need to be commercialized for automotive applications, and technologies must be developed to safely handle accumulated hydrogen when a vehicle is not operating – for example, if a car is parked at an airport for weeks at a time. One option is to use boil-off hydrogen in a fuel cell to generate electricity that can be stored on board in batteries when the vehicle is not operating. Or the hydrogen could be stored in a small metal hydride storage tank or burned by a pilot light.

Refueling challenges

Refueling with liquefied hydrogen is complicated by the fact that refueling equipment must be maintained at super-cold temperatures to prevent the hydrogen from rapidly boiling away as it passes through the equipment and into the vehicle storage tank. Even when the equipment is precooled, boil-off rates of up to 25 percent can occur as the hydrogen passes through less well insulated equipment. Hydrogen released during refueling boil-off can be captured and liquefied again, but this adds cost and inefficiency to the process.

Using liquefied hydrogen in fuel cell engines does not pose any special refueling problems. Cryogenic hydrogen also burns very well in internal combustion engines: the cold increases the thermodynamic efficiency of the engine, thereby increasing its power; it reduces the formation of nitrogen oxides; and it combats the tendency of hydrogen to pre-ignite. However, getting the liquefied hydrogen from the vehicle's storage tank to the engine creates some problems. Because it is a liquid like gasoline, liquefied hydrogen must be pumped to the engine. While this is simple at normal temperatures – every gasoline vehicle on the road has an electronic fuel pump – at cryogenic temperatures, conventional pump parts freeze solid. Researchers in Japan and Germany, among others, have developed specially designed pumps that can operate under cryogenic conditions. The pumps are used in prototype liquid-hydrogen-fueled vehicles that are now on the road.

Chemical Carriers for Hydrogen

A third storage option takes advantage of the fact that many chemicals, including easy-to-handle liquids, contain hydrogen in their molecular structures. Storing these chemicals is, in reality, a way of storing hydrogen, as long as a procedure exists to free the hydrogen by a chemical reaction as it is needed. Hydrogen-rich compounds that have been tested for possible use as molecular containers for hydrogen include natural gas, methanol, cyclohexane, methylcyclohexane, and gasoline itself.

Natural gas and methanol are currently two alternative transportation fuels in their own right. Thousands of vehicles using each fuel are currently operating, and the infrastructure to use each fuel in transportation applications is growing, especially in the case of natural gas. Hydrogen can be produced from natural gas or methanol by a chemical process called steam reforming, in which the hydrogen is separated from the molecular complex. If small steam reforming units were installed in automobiles, hydrogen could be stored on board the vehicles in the form of natural gas or methanol.[20] The process of steam reforming is discussed in more detail in Chapter 6.

Onboard steam reforming has been studied extensively as a hydrogen storage option, and several vehicles, including the Department of Energy-sponsored fuel cell bus program now underway in the United States, use methanol reforming as its hydrogen storage technology (see Chapter 8). One problem, however, is that the weight of the steam reforming equipment and the complexity of this system offsets the advantages gained by storing the fuel using existing natural gas or methanol vehicle storage technology. Also, because only a portion of the natural gas or methanol is actually hydrogen, the weight and volume of the storage tank required is still considerable compared with gasoline. Other issues of concern include energy losses, environmental impact during methanol production, and the high toxicity and corrosivity of methanol.

Moreover, since the late 1980s, Los Alamos National Laboratory, in collaboration with General Motors, has been developing a proton-exchange membrane fuel cell powered by hydrogen produced on board a vehicle from methanol.[21] In late 1994, Chrysler announced a research project to study the use of conventional gasoline as a source of hydrogen for fuel cell powered vehicles; hydrogen would be produced by the partial oxidation of gasoline on board vehicles, a process that Chrysler's early analyses indicate may be more efficient than steam reforming of natural gas or methanol.[22]

Over the long-term, fuel cell technology may be developed that is capable of using methanol or natural gas directly in one reaction vessel, thereby eliminating the need to reform the fuels first to produce free hydrogen. Molten carbonate and solid-oxide fuel cells may be able to accomplish the simultaneous reforming of hydrogen-containing fuels and generation of electricity in one reaction vessel (see Chapter 4). This is possible in part due to the high temperatures and pressures under which these systems operate, conditions which foster the decomposition of hydrogen-containing fuels as well as operation of the fuel cell. If these technologies improve, reformed hydrocarbons may become a more practical hydrogen storage option.

Another liquid hydrogen container that may prove practical, especially in developing countries, is ammonia. With three hydrogen atoms and one nitrogen atom, the molecule carries almost as much hydrogen as natural gas, but without the carbon that contributes to greenhouse gas emissions. Reformed ammonia has already been tested in a variety of transportation applications; for example, to propel an X-15 rocket. A large ammonia-manufacturing industry already exists worldwide because of its use in producing fertilizers. Because there is an infrastructure to deliver ammonia for use in agriculture even in remote areas, hydrogen produced from ammonia may be more viable in these locations than gasoline or any alternative fuel, including hydrogen produced from other sources. The Joint Center for Fuel Cell Vehicles in Golden, Colorado, is currently studying ammonia-based hydrogen vehicle technologies.[23]

Toluene and methylcyclohexane are two other hydrogen-rich liquids that have also been studied as hydrogen carriers. Each chemical contains about 6 percent hydrogen in its molecular structure. When heated under moderate pressure in the presence of a catalyst, these chemicals shed some of their hydrogen as a gas and the remaining liquids are converted to benzene and toluene, respectively. This reaction can be reversed in the presence of hydrogen gas to regenerate cyclohexane and methylcyclohexane.

Like steam reforming, the equipment needed to complete these reactions is heavy and complex, thereby limiting these processes' practicality as onboard hydrogen storage options. Moreover, both benzene and toluene are highly toxic chemicals which require special precautions in handling. The use of liquids to reversibly store and release hydrogen may be more practical as a bulk storage technology for temporarily storing large amounts of hydrogen before delivery to markets. Chemical carriers may, for example, compete well against liquefied hydrogen as a method of shipping hydrogen in oceangoing tankers.

Binding Hydrogen to Solids

As one of the most reactive elements in the universe, hydrogen readily bonds with a variety of solid materials. In one type of bonding, these interactions produce a layer of densely packed hydrogen molecules or atoms bound to the surface of a solid. Bonding hydrogen to solids can greatly increase the amount of hydrogen that can be stored in a given space. That is, more hydrogen can be held by surface bonding in a tank filled with

The hydride storage system on this pickup truck provides a 250-mile driving range. Photo: Hydrogen Consultants, Inc.

solids that have an affinity for hydrogen than could be held as a gas in an empty tank of the same size.

Two types of hydrogen storage technologies take advantage of this tendency of hydrogen to cling to solid particles:

- Hydride technology – which uses various combinations of metals to attract and hold hydrogen.
- Carbon adsorbent systems – which use activated carbons, called adsorbents, to capture and hold hydrogen on their surface. This approach has also been applied to storing natural gas.

Hydride storage systems have been tested in several hydrogen-powered vehicles, mostly in Germany. The Berlin hydrogen vehicle demonstration in the early 1980s (discussed in Chapter 9) used hydride systems. Most of the prototype vehicles produced to date by Daimler-Benz in Stuttgart, Germany, have used hydride fuel storage systems. Although compressed gas systems have been most widely used in the United States, and liquefied hydrogen systems are most common in Japan, a few vehicles equipped with hydride systems have been tested in both countries. Much of the hydride research in the United States has taken place at the Florida Solar Energy Center and at several national laboratories under contract with the United States Department of Energy.

Carbon adsorption technology is still being developed in the laboratory; no prototype hydrogen vehicles operate with this technology.[24] The University of Syracuse's Laboratory for Advanced Storage Systems for

Hydrogen has conducted extensive research into the use of low-temperature carbon adsorbents for hydrogen storage. Although not yet in commercial use, carbon adsorbent systems for natural gas are under study at the Institute of Gas Technology in Chicago, the Atlanta Gas Light Adsorption Research Group, and elsewhere. The development of fully operational natural gas vehicles equipped with a carbon adsorbent fuel storage systems are part of these programs.

Reduced volume, pressure, and cost

The advantage of hydride and carbon adsorbent technologies is that they greatly reduce the volume required to store hydrogen. Compared with the size of compressed storage cylinders containing hydrogen under a pressure of 3,000 pounds per square inch, only about two-thirds the space is needed to store the same amount of hydrogen in solid complexes at 150 psi of pressure. Although this is still more than twice as bulky as liquefied hydrogen systems, and thus nine times bulkier than a gasoline system (see Table 4), this space disadvantage may be offset by the fact that solid complex technologies are much cheaper to build and operate then liquefaction systems.

Main obstacles: Heavy weight and temperature regulation

The key disadvantage of hydrogen storage technologies based on the formation of solid complexes is the weight of the materials used to attract the hydrogen. In most hydride systems developed to date, for example, the hydrogen accounts for only between 1 and 7 percent of the total weight of the storage system; the solid hydride material consumes most of the weight.[25] The total weight of a hydride or carbon adsorption system is twice that of a compressed hydrogen storage system holding the same amount of hydrogen and about four times as great as that of a liquefied hydrogen system (see Table 3).

Researchers are focusing on three different approaches to improve the performance of hydride and carbon adsorbent storage systems: 1) finding solids that are more effective in attracting hydrogen, 2) processing the solid materials used in these systems to increase the surface area available to hold the hydrogen, and 3) adjusting the operating conditions of the system – for example, the temperature and pressure – to maximize the hydrogen storage potential. Processing a solid might entail, for example, pulverizing the solid and contorting each particle so that it is full of tiny pores which can attract and hold hydrogen. The greater the particle surface area, the more hydrogen that can be stored. The third option responds to the phenomenon that, in general, higher temperatures of metal hydrides and lower temperatures of carbon adsorbents increase the capacity of these materials to hold hydrogen.

Temperature regulation during fuel handling is another key issue with this storage technology. The reaction that leads to the adsorption of hydrogen onto the surface of a solid releases heat. Conversely, the hydrogen-solid complex must be heated in order to release the hydrogen and return it to a gaseous state so that it can be used as a fuel in either a combustion engine or a fuel cell engine. Thus, on one hand, refueling requires that the heat produced by adsorption be continually drawn away from the system, while on the other hand, when fuel is needed in the engine, the system must be warmed in order to regenerate gaseous hydrogen. This temperature dependency provides one of the key safety advantages of hydride storage systems. In the event of an accident rupturing the fuel tank, hydride particles spill like a pile of sand: the hydrogen remains bonded to the hydride as nonflammable particles.

To help with temperature management, storage tanks must be equipped with a network of pipes. During refueling, liquid coolant flows through the pipes, cooling the solids and drawing off the heat generated. Conversely, when fuel is being drawn from the tank, warm liquids flow through the pipes to provide enough heat to release a constant flow of gaseous hydrogen. In vehicle applications, waste heat from internal combustion engines is sufficient to keep hydrogen steadily flowing out of the solid complexes as long as a little compressed hydrogen is available to start the engine. If low-temperature fuel cells are used, however, engine heat is insufficient to sustain the flow of fuel, and an onboard heating unit must be used to warm the fuel storage tank during operation, an additional step that lowers the energy efficiency of the engine system.

There are several other technical problems with storage systems based on the formation of hydrogen-solid bonds. Providing adequate passageways through the solid particles so that the hydrogen can reach all the reactive surface areas in the tank is one issue. Contamination of the solids by water or air pollutants that may coat the surface and prevent hydrogen adsorption from taking place poses another problem. A related issue is that not all hydrogen, once adsorbed onto the surface of a solid, is released – even if the solid is extensively heated. The fuel recovery factor often has been found to fall below 80 percent of the fuel initially stored in the system; that is, less than 80 percent of the hydrogen can be extracted from the hydride and used to propel the vehicle. This lowers the usable fuel capacity of the system, which can be countered only by increasing its size and weight. Finally, the storage capability of these systems has been found to deteriorate after a number of refueling cycles has taken place as pores in the solid clog or the solids chemically decompose.

Metal hydride research

Despite the many technological issues facing hydrogen solid-complex storage systems, this technology was widely viewed in the 1970s and 1980s as the most desirable approach to hydrogen storage in transportation applications because hydride storage is achievable at low pressure and moderate temperatures. During that time, most research focused on the development of systems using metal hydrides as the hydrogen-attracting solid.[26]

The first metallic compounds capable of adsorbing and releasing hydrogen were discovered in the late 1960s. Since then a variety of metal complexes have been identified which have the ability to attract and release hydrogen under moderate temperatures and pressures. Lanthanum-nickel alloys and alloys of iron and titanium are two of the complexes that have been most extensively studied in hydrogen storage systems.[27]

Carbon adsorbent research

Various types of nearly pure carbon materials, called activated carbons, have been specially treated to increase their reactivity and used to store gaseous fuels. Interest in using carbon adsorbents to store hydrogen aboard vehicles has grown in the past five years as experiments have shown that hydrogen adsorbent rates at least equal to those in metal hydride systems, and in some cases higher, can be achieved with carbon adsorbents, often at a lower cost. Unfortunately, the best adsorption has been found to occur at temperatures several hundred degrees below zero Fahrenheit. At -200 degrees Fahrenheit, hydrogen adsorption by activated carbons has been shown to increase by a factor of five compared with adsorption rates at room temperature; a fifteenfold increase in fuel storage occurs when temperatures are dropped to -375 degrees Fahrenheit.

Sponge Iron: Rusting to Release Hydrogen

The rusting of iron provides the basis of another hydrogen storage system now under development.[28] This system takes advantage of the chemical oxidation of iron by water, commonly known as rusting, in which the oxygen in water reacts with iron to form ferric oxide, or rust, releasing hydrogen as a by-product. Rusting is a reversible process: if rusted iron is heated in the presence of hydrogen, water is regenerated and the iron is restored.

The quantities of hydrogen produced naturally by the rusting of materials in the world are too dispersed to be of interest as a source of hydrogen. However, systems have been developed recently in which the rate of rusting is dramatically accelerated by heating the water to 1,500 degrees Fahrenheit and exposing this superheated steam to pulverized sponge iron. Sponge iron is a form of manufactured iron containing a large number of pores, like a sponge, that increase the total surface area of the solid. This large surface allows ample contact between the iron and water so that a high rate of rusting can occur. Coating the iron with catalysts can lower the reaction temperature to about 400 degrees Fahrenheit and accelerate the process of rusting and hydrogen generation.

Interest in sponge-iron storage technology for hydrogen vehicles has surfaced only in the 1990s. Although not yet demonstrated in a hydrogen vehicle, technical and economic analyses suggest that it may be competitive with other hydrogen storage options. H Power Corp. in New Jersey is currently a leading developer of sponge-iron technology.

Two challenges: Heavy weight and iron replenishment

About one gallon of water is needed in this system to produce one pound of hydrogen containing the energy of about half a gallon of gasoline. Thus, cars that "run on water" must carry a water tank only about twice as large as a gasoline tank. The water is regenerated when hydrogen is used as a fuel.

However, water-powered cars must carry iron as well as water. Replenishing and regenerating iron presents the key challenge for this technology in vehicle applications. One answer is simply to exchange the container of rusted iron at each refueling stop with a container containing fresh iron. The rusted iron could then be regenerated at some location where hydrogen is available to reverse the rusting process and regenerate the iron.

The weight and size of the iron-filled reaction chamber poses another problem. The amount of sponge iron needed on board to give a vehicle a driving range equivalent to a gasoline vehicle could weigh as much as 1,000 pounds. Although this is heavier than a gasoline system, it is only half the weight of a comparable metal hydride storage system and less than 20 percent of the weight of a lead-acid battery pack.

Notes

1. F. Wyczalek, "European, American and Japanese Electric Vehicles," *Proceedings of the 26th ISATA Conference* (Aachen, Germany, September 13-17, 1993).
2. "GM's Electrifying Two-Seater," *U.S. News and World Report*, January 15, 1990.
3. "Nation's Capital to Celebrate Earth Day with Environmentally-Friendly Buses," *Alternative Energy Network Online Today*, April 18, 1994.
4. James MacKenzie, *The Keys to the Car* (World Resources Institute, Washington, DC, 1994), p. 42.
5. D. Illman, "Automakers Move Toward New Generation of 'Greener' Vehicles," *Chemical & Engineering News*, August 1, 1994.
6. "Appendix A: Energy Pathways," *Hydrogen Program Plan FY 1993 – FY 1994* (United States Department of Energy, Golden, CO, 1992).
7. J. Tachtler and A. Szyszka, "Car Fueling with Liquid Hydrogen (Neunburg vorm Wald) Solar Hydrogen Project," *International Journal of Hydrogen Energy,* Vol. 19, No. 4, April 1994.
8. M. Wald, "A Quick Charge for Electric Vehicles," *The New York Times*, February 16, 1994.

9. MacKenzie, *The Keys to the Car, op. cit.,* pp. 42-50.
10. A number of reports examining hydrogen storage technologies provide much of the information contained in this chapter. In addition to the citations for specific facts which appear below, several of the key references include: United States Department of Energy, "Feasibility of OnBoard Hydrogen Storage for Fuel Cell Vehicles," (Washington, DC, January 1993); United States Department of Energy, *Hydrogen Program Plan FY 1993 – FY 1994, op. cit.*; Jeffrey Bentley, *Hydrogen Storage for Transportation Fuel Cells* (A.D. Little Co., Cambridge, MA, 1993); and H. Buchner, "Hydrogen Use – Transportation Fuel," *International Journal of Hydrogen Energy*, Vol. 9, No. 6, June 1984.
11. James S. Cannon, *Paving the Way to Natural Gas Vehicles* (INFORM, Inc., New York, 1993).
12. "Big Three Form NGV Tank Alliance," *Natural Gas Fuels*, August 1994; and "Big Three Form Natural Gas Vehicle Technology Consortium," *Clean Fuels Report*, September 1994.
13. "Standardized CNG Pressure: 3,000 or 3,600 PSI," *NGV Coalition Issue Paper* (Natural Gas Vehicle Coalition, Arlington, VA), not dated.
14. "Economics of 3,600 PSI Vehicle Tank Pressure Delineated," *Clean Fuels Report*, April 1993.
15. B. Duret and A. Saudin, "Microspheres for On-Board Hydrogen Storage," *International Journal of Hydrogen Energy*, Vol. 19, No. 9, September 1994; and "Microspheres, Waste-to-H_2 Debut at Program Review," *Hydrogen Letter*, May 1994.
16. "Liquefied Natural Gas Fuels," special supplement to *Natural Gas Fuels*, September 1994; and "CNG Vehicle Projects Listed," *Clean Fuels Report*, November 1992.
17. Walter Peschka, a scientist at the German Research Establishment, is one of the leading authorities on use of liquefied hydrogen as an automotive fuel. Two articles written by Dr. Peschka published in the *International Journal of Hydrogen Energy* are: "The Status of Handling and Storage Techniques for Liquid Hydrogen in Motor Vehicles," Vol. 12, No. 11, November 1987; and "Hydrogen Combustion in Tomorrow's Energy Technology," Vol. 12, No. 7, July 1988. Another good article by Peschka is "Cryogenic Fuel Technology and Elements of Automotive Vehicle Propulsion Systems," *Advances in Cryogenic Engineering*, Vol. 37, Part B (Plenum Press, New York, 1992).
18. S. Hynek and W. Fuller, "Hydrogen Storage Within the Infrastructure," *Hydrogen Energy Progress X: Proceedings of the 10th World Hydrogen Energy Conference* (Cocoa Beach, FL, June 20-24, 1994).

19. "Magnetic Liquefaction for Hydrogen Under Development," *Clean Fuels Report*, April 1991; and "Magnetic Liquefaction Could Lower the Cost of Liquid Hydrogen," *Clean Fuels Report*, November 1991.

20. J. Amphlett *et al.*, "Hydrogen Production by Steam Reforming of Methanol for Polymer Electrolyte Fuel Cells," *International Journal of Hydrogen Energy*, Vol. 19, No. 2, February 1994; and Y. Jamal and M. Wyszynski, "On-Board Generation of Hydrogen-Rich Gaseous Fuels – A Review," *International Journal of Hydrogen Energy*, Vol. 19, No. 7, July 1994.

21. Interviews at and tour of the Los Alamos National Laboratory fuel cell lab, December 5, 1994.

22. "Chrysler: Liquid Hydrocarbon for Car Fuel Cells," *The Hydrogen Letter*, January 1995.

23. A. Miller, "Ammonia Fuel Cell Vehicles: Fuel and Power Source of the Future" (Joint Center for Fuel Cell Vehicles, Golden, CO, May 1995); and interview with Arnold Miller, Director, Joint Center for Fuel Cell Vehicles, May 15, 1995.

24. Some articles about carbon adsorption include: "Activated Carbon Storage Looks Good," *The Hydrogen Letter*, July 1993; numerous papers by James Schwarz of the Laboratory for Advanced Storage Systems for Hydrogen, including presentations in the proceedings of the 3rd and 4th *National Hydrogen Conferences* in 1992 and 1993; and R. Chahine and T. Bose, "Low-Pressure Adsorption Storage of Hydrogen," *International Journal of Hydrogen Energy*, Vol. 19, No. 2, February 1994.

25. Y. Shmal'ko *et al.*, "Use of Metal Hydrides in Systems for Supplying Vacuum Physical-Energy Installations," *Hydrogen Energy Progress X: Proceedings of the 10th World Hydrogen Energy Conference* (Cocoa Beach, FL, June 20-24, 1994).

26. Some references on hydride technology include: J. Reilly and G. Sandrock, "Hydrogen Storage in Metal Hydrides," *Scientific American*, February 1980; S. Suda, "Metal Hydrides," *International Journal of Hydrogen Energy*, Vol. 12, No. 5, May 1987; O. Bermauer, "Metal Hydride Technology," *Ibid.*, Vol. 13, No. 3, March 1988; and 31 papers on hydrides published as part of the *Hydrogen Energy Progress X: Proceedings of the 10th World Hydrogen Energy Conference* (Cocoa Beach, FL, June 20-24, 1994).

27. L. Ming *et al.*, "The Hydriding and Dehydriding Kinetics of Some RCO5 Alloys," *Hydrogen Energy Progress X: Proceedings of the 10th World Hydrogen Energy Conference* (Cocoa Beach, FL, June 20-24, 1994).

28. "Hydrogen Cars Pump Iron," *The Economist*, May 16, 1992; and J. Maceda and John Wills, "Advanced Hydrogen Generation and Storage Systems for Fuel Cell Vehicle Support," presented at *Windsor Workshop on Alternative Fuels* (Toronto, Canada, June 1992).

Chapter 6

Hydrogen Production and Distribution

Hydrogen vehicles cannot be used widely without an infrastructure that can both produce large quantities of hydrogen in a cost-effective manner and transport the fuel to market. Technologies available today can produce, distribute, and store hydrogen: The fuel can be made easily from natural gas and, if blended with natural gas, transported in existing natural gas pipelines. But other demonstrated methods of producing hydrogen, including production using renewable resources, are not yet widely commercialized because of high costs and lingering technical problems. Thus, a sustainable transportation system based on hydrogen produced from renewable energy resources faces very significant technological and economic hurdles.

Hydrogen distribution poses a more straightforward challenge: The key obstacle here is the sheer size of the distribution network that must be established in order to make hydrogen fuel widely available. A distribution infrastructure for pure hydrogen now exists but it is tiny, limited to short stretches of pipeline and to small markets served by specially designed truck, rail, and barge tankers. Similarly, current bulk hydrogen storage systems are also limited to a few operations with small capacities.

In both production and distribution, the growing use of natural gas as a vehicle fuel should facilitate the entry of hydrogen into the energy marketplace.

Linking Hydrogen and Natural Gas

Current production and distribution systems for hydrogen, and those under development, are directly analogous to corresponding systems now in place for natural gas. The connection is most obvious in the area of hydrogen production. Most of the hydrogen produced today is generated by the thermochemical conversion of natural gas. Because this is the cheapest and most established method of generating hydrogen, natural gas is likely to remain a key feedstock for hydrogen until production technologies based on renewable energy resources are commercially competitive.

Similarly, in fuel distribution, pipeline transportation technology for natural gas and hydrogen are very similar. In fact, the 1.3-million-mile natural gas pipeline system in place today in the United States can carry significant amounts of hydrogen blended with natural gas without any significant modification. As hydrogen begins to replace natural gas as an energy resource, natural gas pipelines could be modified to carry hydrogen safely, and new compression equipment and pipeline sizes could be added to compensate for differences in the physical characteristics of the fuels. Some components of the natural gas pipeline system would be transferable to hydrogen pipelines, thereby lowering the cost of hydrogen pipelines compared with systems built from scratch.

Perhaps the most important contribution that the natural gas pipeline infrastructure will make to ease the introduction of hydrogen is the pipeline rights-of-way that now connect natural gas-producing regions with consumers around the country. These rights-of-way should be directly usable by a hydrogen infrastructure, even if the natural gas pipelines themselves must be replaced with systems designed specifically for hydrogen.

The rights-of-way of natural gas pipelines may prove to be especially compatible with a hydrogen infrastructure because many of the prime natural gas-producing areas of the United States are also areas that receive high levels of solar energy. The four largest natural gas-producing states – Texas, Louisiana, Oklahoma, and New Mexico – account for 79 percent of the country's natural gas production.[1] All also receive more sunshine than the national average and, in most places, receive adequate rainfall to support large-scale hydrogen production. As natural gas reservoirs gradually become depleted, much of the overlying land surface will be suitable for hydrogen-manufacturing plants using solar technologies. The pipeline system, or at least the rights-of-way, will already be in place to transport the hydrogen to market centers.

There is also significant overlap exist between transportation of natural gas and of hydrogen in a liquefied form. In fact, the current international standards for marine shipment of both fuels are identical. The experience gained by the transporters of liquefied natural gas would be an invaluable asset as the market for shipping liquefied hydrogen grows.

Hydrogen Production Technologies

Given high enough temperatures and pressures, the right catalysts, sufficient space, and adequate money, hydrogen fuel can be produced today from virtually any compound that contains hydrogen atoms in its molecular structure. In order for hydrogen to emerge as the key energy carrier in a sustainable energy economy, however, hydrogen fuel must ultimately be produced from a renewable energy resource in a manner that produces no significant pollution. Moreover, it must be done cost-effectively and use materials that are widely available. It must also be logistically viable; it is impractical, for example, to put wind farms on top of every mountain in the Himalayas just because the wind resources there are abundant. Finally, until economical renewable energy technologies are widely available, environmentally sound alternative hydrogen production pathways using conventional resources, such as natural gas, must provide feasible, transitional methods of obtaining hydrogen.

All of the methods of producing hydrogen that are practical today (discussed below) suffer from distinct drawbacks, mainly economic and technical in nature. Each method, however, is now or could within a decade play a vital role in the production of commercial quantities of hydrogen.[2]

Summary of Hydrogen Production Technologies

Converting Fossil Fuels into Hydrogen

Thermochemical conversion of fossil fuels, predominantly natural gas and coal, into hydrogen through steam reforming, partial oxidation, or pyrolysis.

Advantage: Technology using natural gas as a feedstock is currently the most highly developed and least expensive method of producing hydrogen.
Disadvantage: Although practical for small-scale uses of hydrogen, reliance on fossil fuels is not sustainable. More expensive and less efficient than burning fossil fuel directly. Large environmental impact results when coal is used as a feedstock.

Using Electricity to Produce Hydrogen

Electrochemical processes use electricity to produce hydrogen, usually by splitting the water molecule through electrolysis.

Advantage: Technology is highly developed and can be powered by sustainable sources of electrical generation, such as hydropower, direct solar energy, wind, geothermal energy, and biomass burning.
Disadvantage: Expensive because must produce electricity first. Most current applications of electrolysis use electricity from nonrenewable resources.

Producing Hydrogen Directly from Renewable Sources

1) Thermochemical Conversion of Biomass: Turning organic matter into hydrogen through partial oxidation, pyrolysis, and steam reforming.

Advantage: Biomass is abundant and renewable.
Disadvantage: Need more material compared with fossil fuels and costly equipment to obtain same amount of hydrogen, thus not cost-competitive with fossil fuel feedstocks. Biomass production requires large land areas and biomass itself has a low energy density.

2) Photochemical Conversion: Using direct sunlight to heat catalysts that trigger the splitting of the water molecule into hydrogen and oxygen.

Advantage: Achieves electrolysis without a separate stage to generate electricity.
Disadvantage: Systems are in early stages of development and are very inefficient.

3) Photobiological Conversion: Collecting hydrogen generated as a waste product by some strains of algae and bacteria.

Advantage: Relies on a natural process.
Disadvantage: Laboratory systems that use these organisms are in the very early stages of development and are extremely inefficient.

4) Thermal Decomposition: Heating water to more than 5,600 degrees Fahrenheit to break molecular bonds and make it decompose into its constituent parts.

Advantage: Research indicates that solar furnaces using focused sunlight may achieve these super-hot temperatures.
Disadvantage: Comparable temperatures have only been achieved in nuclear power plants. Large heat input makes the process wasteful.

Thermochemical Processes: Converting Fossil Fuels into Hydrogen

More than 99 percent of the hydrogen produced in the world today is obtained from fossil fuels.[3] While not a long-term source of hydrogen in a sustainable energy economy, hydrogen produced thermochemically using fossil fuels as feedstocks is a practical source of hydrogen in the early stages of the research, development, and commercialization of hydrogen vehicles. These processes currently provide the least expensive method of producing hydrogen.

Some hydrogen is produced as a by-product during the refining of oil, but natural gas is usually the chemical feedstock for processes that deliberately set out to extract hydrogen from fossil fuels. The prime ingredient in natural gas is the methane molecule, which contains 80 percent hydrogen atoms – a higher percentage than any other fossil fuel on Earth, including molecules of crude oil. Methane consists of one atom of carbon and four atoms of hydrogen. By weight, hydrogen accounts for 25 percent of the typical natural gas molecule, compared with 15 percent of gasoline and less than 10 percent of coal.

The most common chemical process by which hydrogen is stripped away from the carbon in natural gas is called **steam reforming**. In fact, steam reforming has been the source for virtually all the hydrogen used to power hydrogen vehicles to date. This role is likely to continue until the cost of producing hydrogen from renewable energy resources declines. Other technologies that are capable of converting fossil fuels into hydrogen and carbon dioxide include **partial oxidation**, or **pyrolysis**.

These traditional hydrogen production techniques suffer the drawback that they are not sustainable because they use nonrenewable fossil fuels as feedstocks. Since 20 percent to 30 percent of the energy of the fossil fuels is burned during conversion to hydrogen, it is usually more efficient to use the fossil fuel directly in some energy application than it is to convert it first to hydrogen. Moreover, the expense of building and operating the equipment needed to convert fossil fuels to hydrogen means that the resulting hydrogen will always be more expensive than the cost of the fossil fuel from which it was produced. The hydrogen that is now produced by steam reforming of natural gas, for example, is between 2.5 and 4.5 times more expensive than natural gas when produced in very large quantities.

Extracting Hydrogen from Fossil Fuels

Steam reforming is a two-step reaction. In the first step, natural gas is mixed with steam superheated to over 1,200 degrees Fahrenheit (usually by burning additional natural gas), under slightly elevated pressures of 600 pounds per square inch, and in the presence of a metal catalyst. Under these conditions, the natural gas and water react to form carbon monoxide and hydrogen. In the second step, called the shift reaction, the carbon monoxide reacts with more water to form additional amounts of hydrogen and carbon dioxide. The overall reaction, therefore, involves conversion of natural gas and water into hydrogen and carbon dioxide. The reaction requires energy: generally between 20 percent and 30 percent of the energy in the natural gas is lost in the conversion to hydrogen.[4] Efficiencies are even lower in small, older reforming units, but could be higher, nearly 98 percent, in larger state-of-the-art systems now being developed.

In partial oxidation, or pyrolysis, fossil fuels are first heated in the presence of limited quantities of oxygen. Instead of burning completely, the hydrocarbons partially decompose into carbon monoxide, hydrogen, and other by-products. More hydrogen is then produced by the shift reaction involving the carbon monoxide and water. Many coal gasification processes use partial oxidation or pyrolysis techniques to turn solid coal into mixtures of carbon monoxide and methane; this methane can, in turn, be converted into hydrogen by steam reforming. Similarly, coal or natural gas can be liquefied by partial oxidation into methanol, which can be steam-reformed into hydrogen.[5]

Nonetheless, thermochemical technologies that use fossil fuels as feedstocks are by far the cheapest method of producing hydrogen. Hydrogen produced by steam reforming of natural gas sold in 1994 on the open market for about $9 per million Btu of energy.[6] Although more expensive than natural gas, it compares much more favorably with the price of gasoline. At the end of 1994, gasoline sold at a wholesale price of 65¢ per gallon, which has an equivalent cost of $5.70 per million Btu. Thus, based on fuel cost alone, hydrogen produced by steam reforming is less than twice as expensive as gasoline today. Other methods of producing hydrogen do not compare as favorably, often costing more than $20 per million Btu of hydrogen energy (see Chapter 7).

Using Electricity to Produce Hydrogen

Although electrochemically produced hydrogen now accounts for less than 1 percent of world hydrogen production, it is the only other method besides steam reforming to provide any hydrogen from technologies now widely used in commercial applications. Electrochemical approaches to producing hydrogen all involve two steps. The first is the generation of electricity. The second step is the use of electricity to produce hydrogen, usually by splitting the water molecule through electrolysis. Like hydrogen produced from reforming of fossil fuels, use of conventional sources of electricity powered by nonrenewable fuels provides a relatively easy way to produce hydrogen needed in the early stages of development and demonstration of hydrogen-powered transportation technologies.

Since electrochemical technologies first generate electricity and then use its energy to produce hydrogen, the cost of the hydrogen produced this way is always going to be greater than the cost of the electricity. Because electricity is currently one of the most expensive forms of energy, electrochemical methods of producing hydrogen are also among the most expensive. The average price of electricity paid by consumers in the United States is about 8¢ per kilowatt hour. This price is equivalent to over $23 per million Btu of electrical energy, about 10 times higher than the wellhead price of natural gas and nearly four times higher than the wholesale price of gasoline. The high price of hydrogen produced from electricity is driven even higher due to energy losses during hydrogen production and to the cost of building and operating the hydrogen production equipment.

Despite its high cost, electrolysis of water is used today in industrial applications to obtain small quantities of hydrogen. Convenience, not cost, is the reason. Electricity is readily obtained from a wall socket and water can be taken directly from a faucet. Pure oxygen needed for steel production and medical purposes, for example, is commonly produced by the electrolysis of water. Electrolysis has become a well-developed commercial technology, and electrochemical approaches, together with steam reforming of natural gas, are the two methods most likely to be widely used to produce hydrogen as the market develops. Indeed, a few prototype hydrogen production facilities now operate using electrolysis.[7]

The most widely used technology to produce hydrogen electrochemically from water is **alkaline water electrolysis**. This system uses several of the principles of alkaline fuel cells discussed in Chapter 4, only it operates in reverse, using electricity to produce hydrogen. The technology of alkaline water electrolysis is well developed, but its applications largely have been limited to small-scale, intermittent operations. Research is needed to improve efficiencies and reduce costs by reducing electrical resistance within the units and by developing new electrodes and membranes capable of withstanding constant, long-term use. Moreover, techniques to ensure the complete separation of the oxygen and hydrogen must be perfected in order to prevent combustible gas mixtures from forming.[8] Alter-

Releasing Hydrogen through Alkaline Electrolysis

To produce hydrogen, electrodes are immersed in an electrolyte solution of potassium hydroxide. The electrolyte solution is divided into two compartments by a membrane between the electrodes. The membrane, made, for example, out of asbestos, permits the flow of electrons and ions between the electrodes, but it is impermeable to gaseous oxygen and hydrogen. As electric current passes through this system, water molecules are split: Hydrogen gas accumulates at the anode electrode and pure oxygen accumulates at the cathode.

An individual electrolysis unit consisting of electrodes, electrolytic solutions, membranes, and associated wiring is called a cell. Each cell is capable of producing hydrogen at only a very slow rate, so hundreds of cells are grouped together either in series, called bipolar cells, or in parallel, called unipolar cells, in order to make a system capable of producing commercial quantities of hydrogen.

natives to alkaline electrolyzers are currently under development. Two promising technologies are the proton-exchange membrane electrolyzer and the high-temperature steam electrolyzer.

Nearly all current applications of electrolysis, including those few instances in which hydrogen is the product, use electricity generated by conventional power plants. Currently, nonrenewable electrical generation accounts for about 81 percent of the electricity generated worldwide; the fossil fuels contribute about 64 percent and nuclear power adds about 17 percent.[9] Thus, when conventional electricity sources are coupled with electrolysis, the hydrogen produced for the most part is not part of a sustainable energy system.

Renewable resources for electrical generation

In the long term, electrochemical hydrogen production relying on electricity generated from renewable energy resources offers a hydrogen production route that is sustainable. These technologies exploit the following resources to generate electricity:

- Hydropower
- Direct solar energy
- Wind
- Biomass
- Geothermal energy.

Only about one-seventh of the world's electricity is produced using renewable electrical generation and most of this electricity is generated from only one renewable energy resource – hydropower. However, other renewable electrical generation technologies are developing quickly and are beginning to enter the commercial marketplace. According to a 1993 study of energy use in the United States since the 1973 oil embargo, electrical generation capacity from renewable solar, wind, geothermal, and biomass resources has grown twentyfold in 20 years, from 500 megawatts in 1973 to 10,000 megawatts in 1992. At the same time, total electrical generation capacity has increased by only 60 percent.[10] This section describes the key renewable electrical generation technologies, in the approximate order of their commercialization in the world today.

Hydropower stations generate about 19 percent of the electricity produced in the world, making hydropower by far the largest renewable energy resource used to generate electricity.[11] Hydropower fuels the nation's largest power plant, the 3,500-megawatt hydroelectric turbine at Grand Coulee Dam in the Northwest, and it forms the backbone of the generating capacity of the Tennessee Valley Authority. The impact of water falling under the influence of gravity is used at hydropower facilities to turn electrical-generating turbines.

Hydroelectric generating plants are certainly not without environmental impacts. The artificial lakes created behind the dams feeding water into hydropower plants raise important land use and water resource management questions. Nevertheless, the generation of electricity with hydropower produces no air pollution, and the resource itself is renewable, as rain and snow return water to the reservoir, from which it falls with the force of gravity.

Hydropower is the proposed source of electricity for several of the largest hydrogen production projects in the world today. The "Euro-Quebec Hydro-Hydrogen Pilot Project," for example, proposes to use hydropower generated in Quebec to produce hydrogen electrolytically. The hydrogen would be transported by oceangoing tanker to Europe, where it would be tested in a variety of commercial applications, including urban buses and aircraft. The "Norwegian Hydro Energy in Germany" project proposes to transport for use in Germany hydrogen produced electrolytically from electricity generated at hydropower plants in Norway. (See Chapter 9 for more details about these projects.) Adding to the economic attractiveness of such projects is the fact that hydrogen can be produced using electricity generated at hydropower facilities during the night, when demand for electricity, and thus rates, are lower.

Photovoltaic cells capitalize on the phenomenon called the photoelectric effect – in which light falling on a metal surface causes it to give out electrons – to produce electricity directly from the energy contained in photons, or units of sunshine. Energy from the sun strikes the photovoltaic cells, temporarily freeing the electrons in the cell's material from their molecular structures and permitting them to move. As electrons move they bump into other electrons, transferring their energy to the second electron in a manner similar to the collision of billiard balls. A photovoltaic cell organizes this transfer of energy between electrons in one direction, thereby creating an electric current.

Moving Electrons in a Photovoltaic Cell

There are no moving parts in a photovoltaic cell. The movement of electrons occurs within solid layers of chemicals that demonstrate the photoelectric effect when exposed to sunlight. Most photovoltaic cells manufactured to date used ultra-pure silicon with small amounts of other chemicals, such as phosphorus and boron, mixed with the silicon in a process called doping. Layers of silicon doped with different chemicals are layered in a way that promotes the organized flow of electrical energy from the side of the cell exposed to sunlight to the other side, where it is tapped as electricity. Silicon cells remain expensive because of the cost of purification, doping, and cell manufacturing. New photovoltaic cells using thin films, composed of mixtures of cadmium and telluride or gallium and arsenide instead of traditional doped silicon, are attracting interest.

Several test facilities currently use photovoltaic cells to power electrolyzers that split water to create hydrogen. The largest such project to date is the 270-kilowatt solar photovoltaic array and electrolysis system in Nuenburg vorm Wald, Germany. The next largest project is the HySolar Project, cosponsored by Germany and Saudi Arabia, which includes a 100-kilowatt and a 2-kilowatt photovoltaic pilot plant located in Riyadh and Jeddah, Saudi Arabia, respectively, and a 10-kilowatt facility operating in Stuttgart, Germany. Other projects include a Spanish energy research facility in Huelva, Andalucia, which runs an 8.5-kilowatt photovoltaic array connected to a hydrogen electrolyzer; a 9.2-kilowatt photovoltaic array and electrolyzer at the Schatz Energy Research Center in Arcata, California; and a 3.5-kilowatt system at the University of California's Riverside campus.

Although still expensive, tremendous strides have been taken in reducing the cost of photovoltaic cells. The average price of silicon solar cells in 1970 was $3,000 per watt (in 1993 dollars); by 1980, the wholesale cost had dropped to $21 per watt. Solar cells now cost about $4 per watt, a drop of more than 99 percent since 1970. As a result, the global market in solar photovoltaic cells has grown more than tenfold in the last decade, to more than 60 megawatts per year.[12]

Despite the impressive cost cuts, however, photovoltaic cells still produce electricity that is about 6 to 8 times more expensive than electricity generated at conventional coal-burning power plants. Low energy conversion efficiency poses one of the biggest technological problems, because it means more cells are needed – increasing the cost of cell arrays and the space they require. The best efficiency of a cell now in commercial production is about 10 percent, meaning that about 10 percent of the energy contained in the sunlight is converted into electricity, although efficiencies greater than 20 percent have been obtained in some experimental photovoltaic cells.[13] The highest efficiency achieved to date is 33 percent. As a result, applications of solar cells are still largely limited to retail products such as calculators and wristwatches and to providing electricity in remote locations where the cost of extending transmission lines is prohibitive.

Solar-thermal electrical generating systems are of three general types, all of which have been demonstrated as technologically viable and are in the early stages of commercial application. These systems – central receivers, dish systems, and troughs – have in common the concentration of sunlight by lenses and mirrors to create a point of intense heat which can be applied to generate electricity using conventional power boilers and turbines.[14] The space requirements for solar thermal technologies are significant: About one square mile is needed for each 100 megawatts of electrical generating capacity. Except for the space requirements, which do not permanently scar the landscape, solar thermal technologies release no air or water pollution. Although very promising for future use as part of a hydrogen energy economy, solar thermal technologies are not now used to power electrolyzers to produce hydrogen.

- Central receivers use hundreds of mirrors, called heliostats, to focus sunlight on a single point. The temperature at this point can easily exceed 1,500 degrees Fahrenheit. In advanced systems, computers are used to adjust the direction and tilt of the heliostats along two axes (up and down and east and west) so that they capture a maximum amount of available light as the sun moves across the sky. The heat generated at the focal point boils a fluid, turning it to steam that in turn generates electricity in a turbine. The largest central receiver system built to date is a 10-megawatt "power tower" built in the early 1980s in California's Mojave Desert. Called Solar One, the plant was tested for several years, but experienced a host of technical problems. A new plant, called Solar Two, is being built at the same location by a consortium of utilities and government agencies at a cost of $35 million. Solar Two will include a molten-salt heat storage system capable of retaining the heat of the sun for use in generating electricity after dark. If successfully demonstrated, the addition of the heat storage system will enable solar generation technologies to operate around the clock as base-load power generators, despite the intermittent nature of sunlight.

- Dish systems consist of small engines or boilers mounted at the focal point of individual, concave mirrors. The heat from each mirror is used by the equipment at the focal point to generate a small amount of electricity. The electricity from several dish systems can be combined to increase the total output of a field containing many dish systems. Hundreds of solar thermal dish systems are now operating worldwide at sizes up to 25 kilowatts each. One of the largest dish projects is in Shenandoah, Georgia: 100 dishes, each measuring 22 feet in diameter, generate a total of 400 kilowatts of power. Dish technology is installed at various locations in California, for a total of 5 megawatts of generating capacity.
- The largest solar thermal power facility in the world uses a trough system. This 354-megawatt Solar Energy Generating Station, composed of nine individual units, was built in Southern California by an Israeli company, LUZ International. A long, semicircular trough lined with a mirror focuses sunlight on a tube suspended along the length of the trough. This is the simplest of the three solar thermal technologies, in part because the position of the mirrors is adjusted along only one axis (up and down) to follow the movement of the

sun. Fluid in the tube is heated and piped to a boiler where electrical generating takes place. LUZ added trough units incrementally between 1984 and 1990 and sold the power produced to Southern California Edison Company, which distributed it to customers in the Los Angeles area. During periods of peak sunshine, the most efficient of the nine LUZ units converted 23 percent of energy of the sunlight into electricity, more than twice the peak efficiency of commercial photovoltaic cells.

Trough technology is the cheapest of the three solar thermal technologies, yet costs are still slightly higher than electricity generated by most conventional technologies. Electricity generated at the LUZ plant, for example, costs more than 10¢ per kilowatt-hour, or about twice the cost of power from conventional coal or natural gas plants. Poor economics led to the closing of the LUZ plant in 1991 after an investment of more than $1.2 billion. It was subsequently sold to another Israeli company, Solel, which has been using the facility to develop and test improved parabolic trough systems and to sell electricity produced by the plant.

Major improvements in **wind** technology during the past decade have dramatically reduced the cost and improved the efficiency of generating electricity from wind resources. About 20,000 wind machines provide about 3,000 megawatts of electrical generation capacity throughout the world today, up tenfold from a decade ago. Wind machines, modern-day windmills, capture wind and use its energy to power a turbine that produces electricity. The largest windmills today are each capable of producing up to 750 kilowatts of electricity, although smaller, 100-kilowatt windmills are more common.

No wind projects currently include hydrogen production as an application of the generated electricity; however, one proposed project, in Palm Desert, California, plans to use wind power to produce hydrogen. More than 90 percent of the world's wind electrical generating capacity is located in California – not because the state has the greatest wind resources, but because of government policies specifically advantageous to renewable energy technologies.More than 90 percent of the wind energy resources in the United States are located in the Great Plains; the potential electric power from wind in just two states, North and South Dakota, exceeds the total electrical generation in the country. Overall, windmills could provide more than four times the country's current total electrical generation.[15]

In addition to being an abundant energy resource, wind power is currently the cheapest and most highly developed renewable resource technology with the exception of hydropower. Wind systems now cost between $1,000 and $1,200 per kilowatt of capacity to build and they produce electricity for about 7¢ per kilowatt hour, only 20 to 40 percent more than power generated from coal or natural gas power plants. One wind project in the Midwest has been designed to produce electricity for just 4¢ per kilowatt hour. The American Wind Energy Association, a trade group, has set a goal of 10,000 megawatts of electrical generation from wind plants in the United States by the year 2000.[16]

Biomass burning is the renewable energy resource most widely used in the world today. About 10 percent of world energy production comes from the various uses of biomass – organic matter that includes forest products and agricultural crops. Most of this is wood burned to generate heat for heating, cooking, and industrial processing. Less than 1 percent of the biomass used as an energy resource today is burned to generate electricity, and none of this electricity is currently used as part of a hydrogen production system. Use of biomass energy in the United States is sparse; only 2.8 quadrillion Btu of energy, about 3 percent of domestic energy consumption, comes from biomass, mostly from burning wood and garbage and from the conversion of crops into ethanol, which is then used as an automotive fuel.[17]

For producing hydrogen, the partial oxidation or pyrolysis of biomass, and the conversion of the gasified products through the shift reaction, probably offers a more economical and environmentally sound production method than burning biomass to generate electricity. This use of biomass, a thermochemical process, is discussed below.

Geothermal energy is another renewable resource that can be used to generate electricity. Geothermal power taps heat within the earth to generate electricity, to heat steam for industrial processes, or to warm buildings;

geothermal power currently provides far less than 1 percent of the world's energy supply.[18] In the United States, geothermal generating capacity is 1,739 megawatts, nearly all of which is located at the Geysers generating facility in Northern California.[19]

Geothermal resources are abundant throughout the world, but sites where underground heat reservoirs are close enough to the surface to be tapped for power are limited and often found in remote areas – necessitating a far-reaching production infrastructure and distribution system. Moreover, exploitation of geothermal resources to generate electricity can produce serious air and water pollution. For example, the water gushing up from a geyser may contain sulfur dioxide or hydrogen sulfide, which then evaporates into the air or is reintroduced to the earth, possibly contaminating underground aquifers. Air pollution can be managed with pollution control equipment, and contaminated waters can be reinjected to reduce surface and groundwater contamination. Some geothermal reservoirs, for example, "hot rock" formations, can be tapped without releasing air pollution or contaminated water; clean water can be injected into the earth, where naturally hot rocks heat it into steam that is released back to the surface.

Finally, there are many **exotic electrical generation technologies** involving renewable energy resources that are not as well developed as the technologies discussed above but that may someday provide electricity for hydrogen production. These technologies include systems that tap the energy of incoming and outgoing tides. Temperature differentials between water layers, especially in tropical reaches of the oceans, have also been exploited by ocean-thermal electric conversion technologies. Fusion power, which merges two hydrogen atoms into an atom of helium in a reaction that takes place in the sun, is another potential source of energy for electrical generation on Earth, but so far it has remained out of the grasp of science. These and other methods of producing electricity are not currently included in any projects involving hydrogen production.

Producing Hydrogen Directly from Renewable Sources

The expense and complexity of producing hydrogen electrochemically through a two-stage procedure has prompted the search for other hydrogen production technologies that use renewable energy resources, but which bypass the initial, separate conversion into electricity. The four renewable hydrogen production technologies discussed in this section – thermochemical conversion of biomass, photochemical conversion, photobiological conversion, or thermal decomposition – all produce hydrogen in one-step processes. However, they are much less developed than the other renewable and nonrenewable technologies described in this chapter. No prototype hydrogen vehicle projects currently underway rely on any of these production technologies, although further research and development may make them a viable part of a renewable energy system.

Thermochemical hydrogen production using biomass

Thermochemical processes can convert biomass, as well as fossil fuels, into hydrogen. Although the use of biomass as a feedstock in thermochemical hydrogen production has the advantage of relying on a renewable energy resource, it is more costly and technologically complex than processes that convert fossil fuels. One problem is that the hydrogen content of biomass is lower than that of fossil fuels. More biomass material, therefore, must be processed with larger and more costly equipment to obtain a significant amount of hydrogen. Purifying the multiple products from the partial oxidation or pyrolysis of biomass is another problem, because the decomposition of biomass, unlike the breakdown of comparatively uniform fossil fuels, produces a diverse amalgam of compounds.

Land use conflicts stemming from the use of large land areas for energy rather than food production are another issue of concern. A 1993 study conducted at Princeton University suggests, however, that biomass farming and hydrogen production on only 3 percent of the America's land area, equivalent to 70 percent of the currently idle cropland in the United States, would supply enough hydrogen to power all the light-duty vehicles expected to be on American roadways in 2010, assuming they are powered with fuel cells.[20]

Despite the obstacles facing use of biomass as a feedstock for thermochemical hydrogen production, interest in this option is growing. In early 1994, a project began at the National Renewable Energy Laboratory in Colorado that will extract the hydrogen in woody plant material by pyrolysis, followed by steam reforming.[21] Another new thermochemical hydrogen production project using switchgrass is underway in Alabama.[22] The high cost of biomass conversion can be reduced by technologies now under development, such as the HYNOL process, which mix biomass with natural gas prior to partial oxidation.[23]

Photochemical conversion: A one-stop hydrogen shop

Photochemical conversion systems are only in the early stages of development, with most research still being performed at the laboratory benchscale. In these systems, sunlight strikes a container holding a liquid electrolyte and either a suspension of metal catalysts mixed in the liquid or surfaces containing fixed semiconductor catalysts immersed in the liquid. In either case, the catalysts absorb the solar energy, creating localized electrical fields that trigger the splitting of water into hydrogen and oxygen. These systems have no moving parts and are much simpler than electrochemical systems because they do not require the generation of electricity and the operation of electrolyzers as separate steps.

Efficiencies of photochemical systems in converting solar energy into energy contained in hydrogen have generally been much less than 10 percent, although efficiencies above 30 percent are possible. One problem is that some of the energy in the sunlight is absorbed as heat by the electrolyte before it strikes the catalysts suspended in the solution. Other issues of concern include catalyst deterioration, corrosion, and the potential for the hydrogen and oxygen gases, which are produced simultaneously, to mix and ignite – presenting a safety hazard.[24] Equipment to physically separate gases adds cost and complexity to photochemical conversion systems.

Photobiological conversion: Bacteria cafeterias

Like direct photochemical conversion technologies, biological hydrogen production techniques are only beginning to be developed, and efficiencies of systems developed to date are typically less than 1 percent. All living organisms are, in a sense, energy factories. They take renewable energy in the form of food or sunlight and, through processes such as digestion (in animals) and photosynthesis (in plants), convert this energy into a form that powers life functions. A few strains of algae and bacteria have been found that produce hydrogen as a waste product; these organisms generate hydrogen in the same way that human beings exhale carbon dioxide, or the way photosynthesis in plants releases oxygen as a by-product.

Enzymes in strains of green algae, including *Chlamydomonas*, have been shown to use the energy of absorbed sunlight to catalyze the splitting of water into hydrogen and oxygen. This reaction tends to occur, however, only at low light intensities, for example at sunset, and under "stressed" anaerobic living conditions, where both oxygen and carbon dioxide are absent. Similarly, strains of photosynthetic bacteria, including *Rhodopseudomonas*, and cyanobacteria, such as *Anabaena variabilis*, also produce hydrogen under certain anaerobic conditions, although they require a continuous supply of metabolic substrates – food – as well as sunlight to complete the reaction.

Development of strains of algae and bacteria that are longer-lived and more efficient in producing hydrogen than these naturally occurring strains is one challenge attracting the attention of genetic engineers. Creating natural environments that are more conducive to hydrogen production by existing strains of algae and bacteria is the subject of research, principally by biochemists.[25]

Thermal decomposition: Making water too hot to handle

Brute force is the technological principal underlying the last method of producing hydrogen. Raised to a high enough temperature, water, like all other chemical compounds, will decompose into its constituent elements, hydrogen and oxygen. Thermal decomposition requires temperatures in excess of 5,600 degrees Fahrenheit, but the burning of fossil fuels, biomass, and other combustible materials are normally capable of generating temperatures no greater than 4,500 degrees Fahrenheit, and generally produce temperatures several thousand degrees lower. The extraordinarily high temperatures needed to trigger the thermal decomposition of water, however, have been achieved primarily in nuclear power plants and in electrically charged plasma torches. Solar furnaces – systems that focus solar rays with mirrors or lenses – also have generated temperatures above 5,000 degrees with focused sunlight, indicating that direct use of solar energy may also provide a way to thermally decompose water.

Interest in thermal decomposition technologies peaked in the 1970s, simultaneously with the height of nuclear power plant construction. As safety, environmental, and economic problems began to plague the nuclear power industry, interest in thermal decomposition as a method of producing hydrogen declined. Another problems with thermal decomposition technology is the tendency of hydrogen and oxygen to recombine spontaneously before they can be separated. Despite these problems, thermal decomposition is still discussed as a potential source of hydrogen in the future, perhaps using heat generated at furnaces powered by solar energy.[26]

Hydrogen Distribution: Getting to the Markets

Once produced, hydrogen must be transported to the people who will use it. Most of the hydrogen produced today is manufactured within or near the oil refineries or chemical plants which now constitute the major hydrogen markets. When produced within the site where it is used, the manufactured hydrogen is referred to as "captive" production; when it is produced off-site by an independent producer, it is called "merchant" production. Like captive producers, merchant hydrogen production facilities are generally located near the industrial facilities that use hydrogen, thereby minimizing the need for transporting the fuel over long distances.

In the few cases today where there is a significant distance between the sites of production and use, pipelines, trucks, and rail tankers provide the most widely used transportation systems. There are four short hydrogen pipelines in the United States, and a truck and rail tanker system has been in use for more than 20 years to deliver hydrogen from various Gulf Coast states to the Kennedy Space Center in Cape Canaveral, Florida.

If hydrogen is to be used as a transportation fuel, a much more extensive and elaborate transportation infrastructure will be needed to move it to distant and widely dispersed markets. The existing hydrogen distribution system includes the basic technology for this infrastructure, and the extensive distribution system already in place for natural gas is likely to be a major asset in the development of a hydrogen transportation infrastructure. The natural gas system in some instances can be used directly to transport hydrogen, with only minor modification. Where technology specific to hydrogen is needed, the existence of analogous systems for natural gas, such as liquefaction technology, may ease the development and implementation of new systems capable of moving hydrogen.

Hydrogen Pipelines

The easiest and most inexpensive way to transport large quantities of gaseous fuels over land is by pipeline. In the United States, a 1.3-million-mile pipeline system has been built to deliver natural gas from its principal points of production, mainly the South Central and Southwestern states, to users throughout the country. In contrast to the extensive infrastructure for natural gas distribution, the hydrogen pipeline system in the United States is a minuscule 450 miles. Although natural gas pipes may be used to transport hydrogen, three technical differences between the two fuels must be addressed for this to be viable: The lower energy density and weight of hydrogen; the greater potential for hydrogen to leak; and the embrittlement of some conventional pipes when exposed to hydrogen.

The natural gas pipeline system has doubled in total mileage since 1960. It is directly connected to more than 50 million customers who currently use natural gas, and it reaches communities where about 96 percent of the US population lives. While the capacity of the pipeline system is about 23 trillion cubic feet per year, it currently carries most of the 19 trillion cubic feet of natural gas consumed in the country annually.[27]

A large natural gas pipeline typically uses a 36-inch-diameter metal or plastic pipe, although pipes as large as 48 inches in diameter are in use. To squeeze as much natural gas through the pipe as possible, the systems operate at pressures up to 1,000 pounds per square inch. Booster stations situated along the pipeline pressurize the gas as needed to keep the pressure throughout the system nearly uniform and close to the desired rate. To carry as much energy in the form of electricity as the energy carried by a large pipeline transporting natural gas would require 10 of the largest overhead transmission lines currently in use. The energy required to operate pipeline systems is very low. Twice as much energy is lost during transmission of electricity through power lines as is used to operate pipelines. Moreover, the construction cost per mile of pipeline is less than one-third that of a mile of electrical transmission line. Thus, as an energy transporter, natural gas pipelines are efficient, inexpensive, and visually unobtrusive compared with electrical transmission lines.[28]

The 450 miles of hydrogen pipeline in the United States is built mostly with pipes of less than 12 inches in diameter. The six pipelines that are over 10 miles in length, and which connect hydrogen production facilities and chemical plants in heavily industrialized areas, account for more than half of the pipeline network. The rest of the hydrogen pipeline infrastructure is made up of shorter links between neighboring production and user sites. Air Products and Chemicals, Inc., operates two hydrogen pipelines, in La Porte, Texas, and Plaquemine, Louisiana, which together stretch 100 miles. Praxair, Inc. operates three pipelines with a combined length of 160 miles: one linking Texas City, Bay Port, and Port Arthur in Texas; one in Whiting, Indiana; and one in Carney's Point, New Jersey. Finally, Big Three Industries Inc. operates a 14-mile hydrogen pipeline near Bayport, Texas. In total, the American hydrogen pipelines carry less than 0.1 trillion cubic feet of hydrogen per year, less than 1 percent of the flow through the natural gas pipeline system.[29]

There are a number of other hydrogen pipelines around the world. The oldest is a 130-mile system built in 1939 in Germany's Ruhr Valley and operated by the Hüls Company. The pipeline links the cities of Cologne, Dusseldorf, Essen, and Dortmund. Hydrogen from four separate hydrogen facilities feeds into the line, whereupon 12 chemical companies along the pipeline's path draw and use it.[30] L'Air Liquide operates the world's longest hydrogen pipeline – a 250-mile line from Antwerp, Belgium, to points in Northern France. The company also owns seven shorter hydrogen pipelines in Europe. There are also several hydrogen pipelines in Canada, including an 18.6-mile high-purity pipeline between Edmonton and Fort

Saskatchewan in Alberta, completed in mid-1995.[31] Moreover, the Canadian Hydrogen Industry Council has prepared a preliminary design of a hydrogen pipeline from the natural gas fields in Alberta to major markets for industrial hydrogen and clean fuels in Southern California.[32]

Comparison of natural gas and hydrogen pipelines

Despite the apparent compatibility between hydrogen and natural gas pipelines, their potential interconnections in future energy supply systems has been the subject of very little technical analysis. The most comprehensive discussion of pipeline compatibility to date is probably contained in a 1972 report, prepared by the Institute of Gas Technology for the American Gas Association, entitled *A Hydrogen-Energy System.*[33] The issue has more recently drawn the attention of Congress, however. Section 2026 of the Energy Policy Act of 1992 requires the Department of Energy to undertake a five-year study including "at least one program to assess the feasibility of existing natural gas pipelines carrying hydrogen gas, including experimentation if needed, with a goal of determining those components of the natural gas distribution system that would have to be modified to carry more than 20 percent hydrogen mixed with natural gas and pure natural gas."

There are several major differences between pipeline technology for natural gas and hydrogen.[34] First, comparably sized pipelines operating at the same pressure can carry, at best, only about 80 percent as much energy in the form of hydrogen compared with natural gas; they generally operate at much slower transport rates. This results from two partially countervailing differences between the fuels. By volume, hydrogen is less energy-dense than natural gas; at the same temperature and pressure, a given volume of hydrogen gas contains one-third the energy as the same volume of natural gas. Therefore, three times the volume of hydrogen fuel must be moved through a pipeline to equal the energy in a given volume of natural gas transported under the same conditions. On the other hand, at the same temperature and pressure, a given volume of hydrogen gas weighs one-third as much as the same volume of natural gas. As a result of its lightness, hydrogen requires less energy to push it through pipelines. It flows roughly three times as fast as natural gas under the same pressure.

These two effects partially offset each other, leading to the practical experience that the hydrogen energy transport rate is much less than the energy transfer rate for natural gas under similar pipeline operating conditions. Actions to increase the rate of hydrogen flow increase pipeline transportation costs. Boosting the number and size of compressor stations to compensate for the lower energy density of hydrogen, for example, can increase compressor capital costs by between 30 percent and 50 percent compared with natural gas. The increased energy demand of larger compressors raises the total transportation cost even higher.

A second difference between natural gas and hydrogen results from the small size of the hydrogen molecule compared with natural gas. Hydrogen has a diffusion coefficient more than three times higher than natural gas: 0.63 versus 0.20 square centimeters per second. In other words, hydrogen leaks more rapidly than natural gas through tiny gaps in pipe fittings, joints, and welds. The potential for leakage through diffusion exists for all gaseous commodities, including natural gas. Because of its especially high diffusion effect, energy losses and safety problems resulting from leaking fuel may prove especially challenging in the case of hydrogen. A safety analysis performed at the University of Miami in 1993 concluded, however, that hydrogen leakage from pipelines would be less likely to ignite than natural gas leaked from the same pipeline operated at the same energy flow rate. In this study, the rapid dispersion of leaked hydrogen into the surrounding air was found to more than compensate for the increased amount of fuel that leaked through the pipe.[35]

While hydrogen leakage from pipelines does not now pose a major safety risk, leakage could become a significant cause of concern if longer hydrogen pipelines operating under higher pressures are built as part of an expanding hydrogen delivery infrastructure. One answer lies in the development of pipeline materials, either for the pipes themselves or as inserts lining a pipeline wall, that are impermeable to hydrogen despite its high diffusion potential.

A final technological issue involves the embrittlement of some conventional pipeline materials when exposed to hydrogen.[36] Most natural gas pipelines cannot be converted to carry pure hydrogen unless this issue is addressed. Hydrogen embrittlement occurs as a result of chemical reactions between hydrogen and metals in the pipeline wall. Over time, hydrogen gradually works its way into the molecular lattice of metals. The result is a deterioration in the pipeline, termed "metal fatigue," evidenced by the development of cracks and pits; metal fatigue can eventually lead to pipe failure or rupture. Embrittlement is a complex and little understood phenomenon that varies with pressure, temperature, the purity of the hydrogen, and the composition of the pipeline.

Many substances can be used safely in hydrogen pipelines, including prestressed concrete, plastics, low-grade steels, which are naturally resistant to embrittlement, and high-strength steels treated with substances called embrittlement inhibitors, for example chromium. Alternatively, conventional pipes can be lined with inserts made of material that is resistant to embrittlement. Another approach involves mixing hydrogen with other gases (for example, carbon monoxide) that have been shown to inhibit embrittlement, probably by coating the walls of the pipeline and preventing access by hydrogen. Whatever method is selected, pipelines built specifically to carry hydrogen must be resistant to embrittlement.

Interestingly, a high concentration of natural gas has been shown to inhibit embrittlement when mixed with hydrogen. Hydrogen concentration as high as 20 percent have been tested in pipelines without evidence of embrittlement. In fact, fuel blends of 20 percent hydrogen and 80 percent natural gas have been successfully transported through natural gas pipelines without the need to perform any modifications to the pipelines or to alter pipeline operating procedures. Gradually mixing small quantities of hydrogen into natural gas pipelines, therefore, might provide an entry pathway for hydrogen into existing markets, while an infrastructure specifically designed for hydrogen is being built. Mixtures of hydrogen and natural gas can be used in most natural gas applications, including vehicles, as long as the concentration of hydrogen remains relatively low.

Liquid Hydrogen Tankers

Pipelines deliver most of America's merchant hydrogen production to markets. But there is a second delivery system that uses barges, tanker trucks, and rail cars to deliver liquefied hydrogen to its main user – the space program – and to a few other users, mostly chemical companies. Pipelines are not practical for liquefied hydrogen transportation because of the difficulty of insulating the large surface area of a long-distance pipeline. Trucks, barges, and rail cars, on the other hand, can be equipped with super-insulated cryogenic tanks capable of keeping the hydrogen cold while the vehicle itself is in motion. Additional refrigeration would be needed to reduce warming during the transfer of the liquefied hydrogen into and out of these tanks. The basic technology of hydrogen liquefaction and storage has been discussed in Chapter 5.

Only six facilities in the United States manufacture liquefied hydrogen, producing about 500,000 gallons annually – less than 2 percent of the nation's total hydrogen production. Praxair, Inc., owns three of these facilities – in New York, Alabama, and California – and Air Products and Chemicals, Inc., owns the other three – in Louisiana, California, and Florida. In contrast, about 100 facilities in the United States and Canada manufacture liquefied natural gas. Annual consumption of liquefied natural gas in the United States is about 915 million gallons. Since the technology to store and transport the liquefied forms of natural gas and hydrogen are similar, natural gas liquefaction sites could be modified in the future to generate and store liquefied hydrogen.

The Space Program: Main consumer of liquefied hydrogen

During takeoff, the Space Shuttle's engines consume 15,000 gallons of liquefied hydrogen per minute. During the first half hour of flight, the shuttle burns an amount of hydrogen fuel equivalent to the amount of fuel used by 1,000 cars driving roundtrip between New York City and Los Angeles.

The inventory for transport vehicles delivering liquefied hydrogen to the space program includes three barges, each capable of holding 250,000 gallons of fuel, five trucks with a capacity of 13,000 gallons each, and four rail cars, each holding 13,000 gallons. Over the past decade more than 200 million gallons of liquefied hydrogen has been transported to space program facilities. The safety record of this fleet has been excellent; there has never been a serious accident with injuries caused by exposure to the fuel.[37]

The liquefied form of hydrogen, like natural gas, is the most conducive to transportation across oceans. Transoceanic pipelines are too costly and technologically complex to be practical. Since liquefaction produces the most energy-dense form of both fuels, cargo space is minimized when either fuel is liquefied. Today, all natural gas commerce between Europe, North and South America, and the Pacific Rim countries of Asia occurs by oceangoing tankers containing natural gas in liquefied form. More than 100 liquefied natural gas tanker ships ply the world's oceans.

Currently there are no oceangoing liquefied hydrogen tanker ships, although they are being designed under the auspices of the Euro-Quebec Hydro-Hydrogen Pilot Project. Liquefied hydrogen has been selected as the preferred method of transporting hydrogen between Canada and Europe as part of the project, and the project's contractors have designed a ship to carry five super-insulated liquefied hydrogen storage tanks. Each tank will hold 9,000 gallons of fuel for a total capacity of 45,000 gallons. The tanks are so well insulated that no boil-off of fuel is expected to occur for at least 50 days. According to the design, the tanks are to be mounted on barges and loaded as a unit onto the ship. When the ocean crossing is completed, each tank and its barge will be unloaded and towed individually to different locations for use or storage.

Marine standards for the transport of liquefied hydrogen and natural gas are identical. In December 1990, the German government ruled that hydrogen was no more dangerous to transport than natural gas; it approved the delivery of liquefied hydrogen to ports in Germany, if the Euro-Quebec Project moves ahead.[38]

Bulk Storage Systems

Fuel production and end-use are operations that are interrelated, but not simultaneous. No matter how closely matched fuel production is to the end-use market for which it is destined, there is occasionally a need to store fuel until needed. When this occurs near the point of production or along the distribution network, it is called bulk storage. A certain amount of bulk storage is needed for any fuel, to provide security in case of a temporary breakdown in production or to respond to fluctuations in demand.

The bulk storage options for compressed hydrogen are likely to be identical to the systems already in place for natural gas: pipelines, natural underground caverns, and high-pressure cylinders or tanks. Insulated cryogenic storage tanks offer viable storage options for both fuels in the liquefied state. An additional option for hydrogen is reversible conversion to hydrogen-rich compounds.

In times of excess fuel production, the pipeline system itself is one form of storage. Pressure can be increased, thereby squeezing more fuel into the pipeline until demand increases and the excess is depleted.

Large natural underground caverns, including those created by the depletion of oil and natural gas fields, present a second storage option. In the United States, about 400 underground caverns, including depleted oil and natural gas reservoirs, depleted aquifers, and natural or depleted salt caverns, are used now to temporarily store natural gas; their combined capacity is nearly 8 trillion cubic feet.[39] Hydrogen's lighter weight and higher potential for leakage raises some concerns that hydrogen might not be suitable for storage in the caverns now holding natural gas. This issue needs to be addressed on a case-by-case basis. However, the national helium storage project in Amarillo, Texas, has for decades maintained underground storage of 30 billion cubic feet of helium, a chemical that is even more subject to leakage than hydrogen. That cavern is the result of a depleted natural gas field.

The third bulk storage option is above ground, either in high-pressure storage cylinders holding compressed hydrogen or in insulated tanks holding liquefied hydrogen. Both of these options are expensive compared with pipeline or underground storage, but they have applications, especially near the end-use site, for relatively small quantities of fuel. The largest bulk storage tank for liquefied hydrogen in the world today is the 850,000-gallon vacuum-insulated spherical tank at the Kennedy Space Center.

Finally, hydrogen can be combined in a chemical reaction with another substance to form a hydrogen-rich compound that can be stored easily as a liquid or a solid in conventional storage tanks. When hydrogen is needed, the reaction is reversed and the hydrogen is once again liberated as a gas. Reversible chemical conversion options, discussed in more detail in Chapter 5, include sponge iron, methanol, and cyclohexane.

Notes

1. United States Energy Information Administration, *Natural Gas Annual 1991* (Washington, DC, 1992, p. 6).
2. The principal sources of information about hydrogen production used in this chapter include: T. Johansson *et al., Renewable Energy: Sources for Fuels and Electricity* (Island Press, Washington, DC, 1993); United States Department of Energy, *The Potential of Renewable Energy: An Interlaboratory White Paper* (Washington, DC, March 1990); United States General Accounting Office, *Electricity Supply: Efforts Under Way to Develop Solar and Wind Energy* (Washington, DC, April 1993); C. Flavin and N. Lenssen, *Powering the Future: Blueprint for a Sustainable Electricity Industry* (Worldwatch Institute, Washington, DC, June 1994); and J. Bockris, *Energy Options: Real Economics and the Solar-Hydrogen System* (Australia and New Zealand Book Co., Auckland, New Zealand, 1980).
3. C. Winter *et al.,* "Hydrogen as an Energy Carrier: What is Known? What Do We Need to Learn?," *International Journal of Hydrogen Energy*, Vol. 15, No. 2, February 1990.
4. Y. Jamal and M. Wyszynski, "On-Board Generation of Hydrogen-Rich Gaseous Fuels – A Review," *International Journal of Hydrogen Energy*, Vol. 19, No. 7, July 1994; and B. Cromarty and C. Hooper, "Increasing the Throughput of an Existing Hydrogen Plant," *Hydrogen Energy Progress X: Proceedings of the 10th World Hydrogen Energy Conference* (Cocoa Beach, FL, June 20-24, 1994).
5. L. Bicelli, "Hydrogen: A Clean Energy Source," *International Journal of Hydrogen Energy*, Vol. 11, No. 9, September 1986; and D. Considine (ed.,) *Energy Technology Handbook* (McGraw-Hill, New York, 1977), p. 4-36; and "Development of a Multi-Fuel POX Reformer Underway at ADL," *Clean Fuels Report*, February 1995.
6. United States Department of Energy, *Hydrogen Program Implementation Plan FY 1994 – FY 1998* (Washington, DC, October 1993), p. 5.
7. Information package sent to INFORM by Electrolyser Corp. Ltd., 1991; and W. Dontz, "Economics and Potential Applications of Electrolytic Hydrogen in the Next Decades," *International Journal of Hydrogen Energy*, Vol. 9, No. 10, October 1984.
8. "Renewable Transportation Fuels: Evaluation of Technology and Environmental Impact," presentation by Matthew Fairlie, Electrolyser Corp. Ltd., at *Second Annual World Car Conference* (Riverside, CA, January 24, 1995).

9. United States Energy Information Administration, *International Energy Annual 1992* (Washington, DC, January 1994), p. 94.

10. Sun Day: A Campaign for a Sustainable Energy Future, "Twenty Years After the Oil Embargo," (Washington, DC, October 1993), p. 5.

11. *Ibid.*, p. 94.

12. C. Flavin and N. Lenssen, *Power Surge: Guide to the Coming Energy Revolution* (W.W. Norton & Co., New York, 1994), pp. 152-163.

13. Luther Skelton, *The Solar-Hydrogen Energy Economy* (Van Nostrand Reinhold Co., New York, 1984), p. 39-52.

14. *Ibid.*, pp. 132-151; and P. Klimas and D. Menicucci, "Solar Thermal Systems – Ready to Compete," (Sandia National Laboratories, Albuquerque, NM, 1990).

15. A. Cavallo *et at.*, "Baseload Wind Power from the Great Plains for Major Electricity Demand Centers" (Center for Energy and Environmental Studies, Princeton University, Princeton, NJ, March 1994).

16. Flavin and Lenssen, *Power Surge..., op. cit.*, pp. 114-131.

17. United States Energy Information Administration, *Annual Energy Review 1993* (Washington, DC, July 1994), pp. 261-267.

18. United States Energy Information Administration, *International Energy Annual 1992, op. cit.*, pp. xiii-xiv.

19. United States Energy Information Administration, *Annual Energy Review 1993, op. cit.*, p. 277.

20. "Renewable Supply of Hydrogen for Transportation Would Require Small Land Area," *Clean Fuels Report*, November 1993.

21. NREL Launches Biomass Pyrolysis Project," *The Hydrogen Letter*, February 1994.

22. "Electro-Farming Comes to Alabama," *The Hydrogen Letter*, April 1994.

23. "Renewable Transportation Fuels: Evaluation of Technology and Environmental Impact," presented by Stefan Unnasch, Acurex Environmental Corp., at *Second Annual World Car Conference* (Riverside, CA, January 24, 1995).

24. The current status of direct photochemical conversion technologies to produce hydrogen was discussed in 10 papers published in the *Hydrogen Energy Progress X: Proceedings of the 10th World Hydrogen Energy Conference* (Cocoa Beach, FL, June 20-24, 1994).

25. "Photobiological Production of Hydrogen in Long-Term Possibility," *Clean Fuels Report,* November 1994. The current status of photochemical conversion technologies to produce hydrogen was discussed in 5 papers published in the *Hydrogen Energy Progress X: Proceedings of the 10th World Hydrogen Energy Conference* (Cocoa Beach, FL, June 20-24, 1994).

26. "NREL Takes New Look at Direct Solar Water Splitting," *The Hydrogen Letter*, October 1994. The current status of thermal decomposition technologies to produce hydrogen was discussed in 11 papers published in the *Hydrogen Energy Progress X: Proceedings of the 10th World Hydrogen Energy Conference* (Cocoa Beach, FL, June 20-24, 1994).

27. American Gas Association, *1993 Gas Facts* (Arlington, VA, 1994).

28. E. Fein, "Hydrogen: An Accommodating Fuel," *International Journal of Hydrogen Energy*, Vol. 10, No. 5, May 1985; and D. Considine, *Energy Technology Handbook* (McGraw-Hill Book Co., New York, 1977), pp. 4-31 to 4-33.

29. R. Moore and D. Nahmias, "Gaseous Hydrogen Markets and Technologies," and M. Kerr, "North American Merchant Hydrogen Infrastructure," papers presented respectively in the first and fourth *Proceedings of the Annual Meeting of the National Hydrogen Association*, Washington, DC, March 1990 and March 1993.

30. "Hüls Hydrogen Pipeline System," information sheet prepared by Hüls Aktiengesellschaft, not dated.

31. "Praxair Wins Hydrogen Contract and Announces Pipeline Project," *Clean Fuels Report*, November 1994.

32. "Hydrogen Transportation Using Present Pipeline Networks," *Alberta Hydrogen Research Program: Annual Review 1991-1992* (Edmonton, Alberta, Canada, 1993).

33. D. Gregory, *A Hydrogen Energy System* (American Gas Association, Arlington, VA, 1973).

34. Considine, *Energy Technology Handbook*, *op. cit.*; and F. Oney *et al.*, "Evaluation of Pipeline Transportation of Hydrogen and Natural Gas Mixtures," *International Journal of Hydrogen Energy*, Vol. 19, No. 10, October 1994.

35. "Hydrogen Pipeline Leaks Would be Less Hazardous than Natural Gas," *Clean Fuels Report*, February 1994.

36. G. Biggiero *et al.*, "Embrittlement due to Hydrogen in Ferritic and Martensitic Structural Steels," *International Journal of Hydrogen Energy*, Vol. 20, No. 2, February 1995.

37. A. Bain, "Space Shuttle Program," paper published in the *Proceedings of the First Annual Meeting of the National Hydrogen Association* (Washington, DC, March 1990).

38. J. Gretz *et al.*, "Status of the Hydro-Hydrogen Pilot Project (EQHHPP)," *International Journal of Hydrogen Energy*, Vol. 19, No. 2, February 1994.

39. American Gas Association, *1993 Gas Facts, op. cit.*

Chapter 7

The Hydrogen Vehicle Fuel Cycle

Efficiency, Economics, The Environment, and Safety

In order for hydrogen to supplant fossil fuels and become the backbone of a sustainable transportation system, it must be an energy resource that can be produced, delivered, and used efficiently. Hydrogen must be affordable to the consumer, and its environmental advantages must be evident throughout the entire cycle of production and use, not just at the tailpipe of a hydrogen-powered vehicle. Finally, hydrogen must be safe, by reasonable stand-alone criteria and in comparison with other fuel options.

Different pathways of hydrogen production, distribution, and use as a transportation fuel bear with them different efficiencies, costs, environmental effects, and safety aspects. These pathways are called hydrogen fuel cycles. Some hydrogen fuel cycles are significantly more energy efficient than the conventional gasoline fuel cycle, particularly those that produce hydrogen from natural gas for use in fuel cells. Hydrogen produced today from renewable energy resources is very expensive, but the cost of producing hydrogen from natural gas is surprisingly close to the price of gasoline. Currently, all hydrogen vehicles are very expensive prototypes; however, within a decade it may be possible to mass-produce hydrogen vehicles with life-cycle costs comparable to the costs of owning a gasoline vehicle – particularly if costs external to the vehicles' "sticker" price are considered.

The environmental benefits of using hydrogen as an automotive fuel, discussed in Chapter 4, are sustained throughout production and distribution, although some hydrogen fuel cycle pathways, mainly those which produce hydrogen from renewable energy resources, are more environmentally attractive than others. With regard to safety, by some measures hydrogen is inherently more dangerous to handle than other fuels; at the same time, hydrogen also enjoys some safety advantages over other fuels, and technology is already addressing the key safety concerns raised by using hydrogen as a transportation fuel.

Fuel Cycle Energy Efficiency

The total amount of energy consumed in moving a vehicle is equal to the sum of the energy that the engine delivers to power the wheels, plus all the energy from the original energy resource that has been lost en route to the engine. Fuel cycle energy efficiency is measured as the percentage of energy delivered to the wheels of a vehicle compared with the total amount of energy expended during the entire process of fuel production, distribution, and use. There are numerous steps in the process where large energy losses occur, and the greater the losses during each step, the lower the overall efficiency of the fuel cycle. Fuel cycle efficiency is a key factor in determining the amount and cost of fuel ultimately needed to propel a vehicle, the total environmental impact of producing and using that fuel, and the overall safety risks of handling the fuel along the way. Hence, developers of alternative transportation fuel systems are all looking to attain high fuel cycle efficiencies.

To take one example, the fuel cycle for gasoline, the predominant fuel in transportation today, involves: crude oil production from wells; transportation by tanker or pipeline to refineries, where oil is chemically processed into gasoline; distribution to refueling stations, primarily by other pipelines and trucks; fueling; and burning in internal combustion engines aboard automobiles. Although there is only one fuel cycle generally used for gasoline, more than one fuel cycle can exist for any particular fuel. In the case of hydrogen, for example, the fuel can be produced by steam-reforming natural gas, by gasifying biomass, or by electrolysis using solar energy as the source of electricity. Once produced, hydrogen can either be burned in an internal combustion engine or used in a fuel cell to provide power onboard a vehicle. Each hydrogen fuel cycle has its own overall energy efficiency associated profile. One difficulty in analyzing the merits of hydrogen as a transportation fuel is that there is a wide range of energy efficiencies among its different fuel cycles. Some are vastly more efficient than the conventional fuel cycle for gasoline, others are less efficient.

The Highly Inefficient Gasoline Fuel Cycle

Alternative fuels competing with gasoline for a place in the automotive industry have a tempting target. The gasoline fuel cycle, dominated by the highly refined but enormously inefficient internal combustion engine, is extremely wasteful of energy. The amazing performance of the modern automobile comes at a cost. Only 12 to 15 percent of the energy contained in crude oil, the original source of gasoline, is delivered as power to turn the wheels of an average car.[1]

- About 15 percent of the energy content of crude oil is used up to pump it from the ground and to provide heat and power to operate oil refineries.
- In the distribution system, about 3 percent of the energy contained in refined gasoline is used to operate the pipelines and to propel the tankers used to get it to market.
- Once gasoline enters the engine, large energy waste occurs. About 80 to 85 percent of the energy released by burning gasoline in the internal combustion engine is lost, mainly as waste engine heat. The remaining energy is converted into the mechanical rotations of the engine's shafts and gears.
- About 15 percent of this mechanical energy is lost, mostly through friction, as the energy passes through the transmission and on to the wheels.

Table 1. Fuel Cycle Energy Efficiencies

Fuel Cycle	Energy Efficiency*
Fuel burned in internal combustion engine	
Gasoline	12 - 15%
Natural gas	9 - 15%
Hydrogen produced from natural gas	8 - 13%
Hydrogen produced from fossil fuel electricity	2 - 4%
Fuel used in fuel cell engine	
Hydrogen produced from natural gas	28 - 30%
Hydrogen produced from photovoltaic cells	15 - 20%
Hydrogen produced from biomass	15 - 32%
Hydrogen produced from fossil fuel electricity	8 - 10%
Electric battery	
Electricity produced from fossil fuels	20 - 29%
Hybrid electric battery/combustion engine vehicle	20 - 25%

**Energy efficiency is the percentage of the total amount of energy expended in the fuel cycle that is delivered to the wheels of the vehicle.*

Source: INFORM calculations based largely on US Department of Energy, *Hydrogen Program Plan: FY1993 – FY1997,* Appendix A.

Comparison of Alternative Fuel and Engine Efficiencies

As shown in Table 1, using hydrogen in a fuel cell is, overall, more efficient than burning it in an internal combustion engine. Similarly, most fuel-cell uses of hydrogen, regardless of how the fuel was produced, are more efficient than the conventional gasoline-burning vehicle.

The first four rows of Table 1 show overall fuel cycle efficiencies of gasoline, natural gas, and hydrogen when burned in internal combustion engines. The low-end estimates for fuel cycle efficiencies of the alternative fuels fall below even the meager 12 to 15 percent efficiency level of the conventional gasoline fuel cycle. Since natural gas does not need to be chemically processed, it avoids the energy losses inherent in the refining step of the gasoline fuel cycle. It also burns somewhat more efficiently in the engine than gasoline. These gains are offset, however, by energy inefficiencies in storing natural gas onboard vehicles. For example, compressing natural gas, the most common storage option, requires far more energy than pumping gasoline into automotive gas tanks. Thus, the overall fuel cycle efficiency of natural gas vehicles is 9 to 15 percent.

Switching from natural gas to hydrogen use in the internal combustion engine entails only a small additional drop in efficiency, to 8 to 13 percent, assuming the hydrogen is produced by steam reforming of natural gas. If the hydrogen is produced by electrolysis using electricity generated at conventional power plants, which are only about 35 percent efficient themselves, the efficiency of the hydrogen fuel cycle drops to a mere 2 to 4 percent.

Electric motors are more efficient

Given the huge energy waste inherent in using the internal combustion engine, significant increases in transportation fuel cycle efficiency cannot occur without shifting to alternative propulsion systems. This is possible by replacing internal combustion engines with electric motors. Electric motors deliver up to 90 percent of the energy they consume to the wheels of a vehicle. Moreover, electric vehicles are able to capture energy released as heat during braking. Energy captured by such "regenerative" braking systems can be stored in batteries and used again to power the motor. One-third of all energy used by a car is lost during braking; regenerative braking can recapture half of this loss.[2] Greater inefficiencies are encountered, however, during the generation and storage of electricity for use in electric motors compared with the analogous steps in the gasoline fuel cycle. Onboard hydrogen storage systems are only about 75 percent efficient in storing energy; the other 25 percent of the energy is used in moving the hydrogen into and out of the storage system. Similarly, only about 70 to 80 percent of the electricity stored in batteries can be recovered; a battery is not capable of absorbing all the energy delivered to it during recharging.

Taking fuel production, distribution, storage, and use together, the net result is that fuel cycles for several types of electric vehicles are more efficient than the conventional gasoline fuel cycle. Table 1 shows that three electric vehicle configurations – hydrogen in fuel-cell-powered electric vehicles, battery-powered vehicles, and hybrid-electric vehicles that pair batteries with internal combustion engines – all exhibit fuel cycles at least twice as efficient as the conventional gasoline fuel cycle. Further technological development promises to increase these figures.

If hydrogen-powered fuel cells are used to generate the electricity to power electric motors, the efficiency of the engine component of the fuel cycle jumps from the 15 to 20 percent common to internal combustion engines to about 50 percent. When coupled to an electric motor and using hydrogen produced by 70-percent-efficient steam reforming of natural gas, hydrogen fuel-cell-powered vehicles can achieve a fuel cycle energy efficiency of 28 to 30 percent. If the hydrogen is produced from renewable energy sources, fuel cycle efficiencies drop, but remain higher than for gasoline. Efficiencies ranging from 15 to 32 percent are possible using hydrogen produced by electrolysis using electricity generated by photovoltaic cells or by gasification of biomass.

Fuel cycle efficiencies involving fuel-cell-powered vehicles fall below the efficiency of the conventional gasoline fuel cycle only when hydrogen is produced by electrolysis using electricity generated at conventional fossil fuel or nuclear power plants. However, if electricity is stored in batteries on board vehicles, rather than converted to hydrogen and reconverted to electricity by fuel cells, higher efficiencies are possible. Fuel cycle efficiencies between 20 and 29 percent are possible with a battery-powered electric vehicle using electricity generated at conventional power plants. Electric hybrid vehicles equipped with internal combustion engines used to generate electricity rather than to provide power directly to the wheels offer fuel cycle efficiencies between 20 and 25 percent.

Much of the inefficiency of the conventional gasoline fuel cycle stems from the need to equip vehicles with internal combustion engines large enough to meet the peak demand for power – for example, during rapid acceleration. Most of the time, only a small fraction of the power produced by the engine is used for propulsion. The most obvious example is when a car is idling at a traffic light or stalled in traffic congestion. About 10 percent of the energy released from burning gasoline is wasted while automobiles idle in traffic.[3] With the exception of providing a small amount of power needed to operate the radio and other accessories, all of the power produced by the engine during these times is wasted. Even at high cruising speeds maintained on highways, less than roughly one-third of the engine's maximum output is being used. During a typical driving routine in urban areas, a conventional car's engine is operating at peak efficiency only 7.6 percent of the time.[4]

Modifying a vehicle to make it an efficient hybrid requires more than adding a battery system: It entails redesigning the internal combustion engine to operate at a steady speed and storing the excess energy in the battery until needed. The electricity generated by such steady-state internal combustion engines can be used either directly to propel a vehicle or it can be stored in a battery for future use. This configuration not only increases the efficiency of the engine; it also decreases the number of batteries needed by the vehicle because a portion of the vehicle's energy demand can be met directly by the generator. Engine efficiencies above 30 percent are possible when internal combustion engines are operated as electricity generators under steady-state conditions, or about twice as high as the efficiency of the internal combustion engines as they are now used in automobiles. This leads to an average overall fuel cycle efficiency above 20 percent for electric hybrids powered by gasoline.[5]

Hybrids can be designed to burn any combustible alternative transportation fuel – for example, natural gas or hydrogen – or they can retain gasoline as a fuel. The exact fuel cycle efficiency of an electric hybrid varies according to which fuel is used, but the fuel cycle efficiency of any fuel is higher when burned in an electric hybrid vehicle compared with its efficiency when burned in a conventional vehicle. Fuel cycle efficiencies above 30 percent have been projected for the future for electric hybrids powered by internal combustion engines burning natural gas or hydrogen.[6]

The Cost of Hydrogen as a Transportation Fuel

Currently, hydrogen is not mass-produced for use as a transportation fuel. There is no hydrogen delivery and refueling infrastructure, and each hydrogen-powered vehicle manufactured to date has been handcrafted for testing purposes. Consequently, economic analyses of the cost of hydrogen and hydrogen-powered vehicles are based on projections of future costs, using assumptions about the rate of technological innovation and the economies that may be achieved through mass production and marketing. This analysis focuses on the economics of hydrogen vehicles as the technology matures over the next decade. The price of early demonstration vehicles will be considerably higher; on the other hand, costs should drop dramatically as mass production begins – a development that is not likely to begin in earnest within 10 years. To get to this stage, investments in hydrogen vehicle technology must appear to be economically prudent, not only for the long term but for the immediate applications for which they are directed.

In looking at the current economic situation of vehicles powered by gasoline and hydrogen, and at how the economics of these vehicles could change over the next decade, five issues must be considered:

- fuel costs
- vehicle costs
- external costs
- economic credits
- life-cycle vehicle operating costs.

Fuel Costs

Consumers pay two key bills resulting directly from fuel choice: 1) the fuel bill paid every time the vehicle is refueled, and 2) the portion of the vehicle purchase price that reflects the cost of the equipment needed to handle and use the fuel, paid once but subject to periodic maintenance costs. The fuel bill includes the cost of fuel production, distribution and storage charges, the price of operating a refueling station, plus profits or taxes that may be tacked on at each of these stages. The vehicle cost includes the price of the fuel storage system, equipment needed to deliver the fuel to the engine, and the cost of the engine itself.

Gasoline-powered vehicles represent the economic target for alternative transportation fuels to beat with regard to both fuel and vehicle costs. Table 2 contains approximations of the cost of gasoline and hydrogen fuels. The table establishes a benchmark "pump" price of gasoline at $1.25 per gallon, which was close to the average price of gasoline sold in the United

States in 1994. Prices could also be expressed in terms of dollars per million Btu of energy. Since a gallon of gasoline contains 115,400 Btu of energy, price per gallon figures can be converted to price per million Btu by multiplying by 8.67.[7] Thus, $1.25 per gallon is approximately equivalent to $10.84 per million Btu of energy. The table presents hydrogen prices in dollars per "equivalent gallon," which is the energy contained in a gallon of gasoline.

The retail price of any fuel incorporates three main components:

1. the wholesale price
2. the costs of distributing the fuel and refueling vehicles
3. taxes.

Wholesale pricing

The current wholesale price of gasoline is about 60¢ per gallon ($5.20/million Btu);[8] pipeline and truck distribution and refueling station costs add another 25¢ per gallon ($2.17/million Btu), and the average combined state and federal taxes amount to nearly 40¢ per gallon ($3.47/million Btu).[9]

Table 2. Fuel Costs of Gasoline vs. Hydrogen

Fuel Cost Component	Gasoline ($/gallon)	Hydrogen ($/equivalent gallon)
Wholesale Fuel Cost	$0.60	$0.80-$4.00
Fuel Distribution/Refueling	$0.25	$0.65-$2.50
Taxes	$0.40	$0.00-$0.40
Total $/gallon	$1.25	$1.45-$6.90

(In constant 1995 dollars)

Hydrogen transportation fuel prices reported in Table 2 assume some technological advancements that are likely to be achieved within a decade and some economies of scale from mass production and distribution. Thus, hydrogen costs are projections of marketplace prices (expressed in 1994 dollars) that exist now, in the case of hydrogen produced from natural gas, or that might reasonably exist by the year 2000, using other methods of production. [10]

The cost of hydrogen depends most significantly on how and from what resource it is produced. The cheaper production cost shown in Table 2 results from steam reforming of natural gas. The current market price of steam reformed hydrogen is about 80¢ per equivalent gallon ($6.94/million Btu), or only about 33 percent more than the current wholesale price of gasoline.

The cost of producing hydrogen from renewable energy resources is much higher than steam reforming. Prices projected for the next decade range from $1.15 ($10/million Btu) to $4 ($30/million Btu) per equivalent gallon, or from twice as expensive to six times the current cost of wholesale gasoline. Among the renewable resource alternatives, producing hydrogen by gasification of biomass or by electrolysis using electricity generated from wind or hydropower resources are currently projected to be the lowest-cost options, less than $2.50 per equivalent gallon (about $20/million Btu). The direct use of solar energy, either by photovoltaic cells or with solar thermal technologies, is likely to be higher – between $2.50 and $4 per equivalent gallon ($20 and $30/million Btu). More advanced hydrogen-production technologies such as direct photochemical conversion and biological conversion are not sufficiently developed to allow price projections.

Distribution costs

The distribution of fuels consists of two major elements: the transportation to refueling stations and the cost of the refueling operation itself. The benchmark for gasoline distribution costs is about 25¢ per gallon. Like production costs, distribution costs for hydrogen projected to be achievable within a decade are likely to range from levels only marginally higher than gasoline costs to several times more expensive. Transporting hydrogen by pipeline could cost, on average, between 35 and 70¢ per equivalent gallon ($3 to $6/million Btu), although the exact cost depends on the distance involved. Refueling station costs could add another 30 to 80¢ per equivalent gallon ($2.50 to $7/million Btu), depending in part on the type of refueling technology. Compressed-hydrogen storage entails the highest refueling costs, mainly due to the energy used in compressors; carbon adsorption refueling is the cheapest. If hydrogen is liquefied before delivery to a vehicle, the cost of liquefaction, estimated to be about $1/gallon ($9/ million Btu), must be added to the refueling cost.

Different onboard fuel storage systems require different types of hydrogen refueling stations. The refueling technologies most likely to be used are high-pressure gas systems serving compressed gas storage cylinders; low-pressure gas systems serving hydride, carbon adsorption, and sponge-iron storage tanks; and liquefied hydrogen systems serving cryogenic storage vessels. Refueling station costs are projected to range from 30¢ per equivalent gallon for low-pressure storage systems to 70¢ per equivalent gallon for a liquefied hydrogen refueling station.

Taxes and retail prices

Currently, federal and state road taxes add about 40¢ to the price of a gallon of gasoline, or about 32 percent of a $1.25 pump price. It seems unlikely that fuel taxes for hydrogen would ever exceed the comparable tax on gasoline. On the other hand, the many advantages of using hydrogen as part of a sustainable transportation system could justify eliminating taxes on hydrogen as a government policy to promote its use, at least in the early stages of development. Table 2, therefore, includes a likely range of hydrogen road taxes from no tax to a tax of 40¢ per equivalent gallon ($0 to $3/million Btu).

The pump price of hydrogen by the end of the decade may thus range from $1.45 per equivalent gallon ($12.50/million Btu), only 16 percent higher than the current price of gasoline, to $6.90 per equivalent gallon ($60/miliion Btu) – about 5.5 times more expensive than the current price of gasoline. Even small improvements in the efficiency of hydrogen vehicles would erase the differential in fuel price for hydrogen, if actual production costs are at the low end of this range.

Moreover, this analysis assumes that the pump price of gasoline remains the same in constant dollars until at least the end of the decade. Should gasoline prices increase, the gap between hydrogen and gasoline prices would diminish. In fact, gasoline prices will almost without a doubt increase at least 10¢ per gallon, just to reflect the increased cost of producing cleaner varieties of fuel to meet stricter incremental standards. For example, under the Clean Air Act of 1990, use of reformulated gasoline, which produces less air pollution than conventional gasoline, is required in all automobiles operating in the nation's ten most polluted cities. The regulation took effect on January 1, 1995, immediately affecting one-third of all motorists.[11] Higher gasoline prices could also result from stricter underground storage tank requirements at refueling stations, tighter oil tanker construction specifications, higher road taxes, and so on. The cost of oil itself could also rise as depletion extracts a toll in the face of rising global demand or in response to political crises reminiscent of the 1973 oil embargo.

Fuel cells offer hydrogen retail advantage

Barring a major jump in gasoline prices, an equivalent gallon of hydrogen is likely to cost more than a gallon of gasoline to the consumer driving up to a refueling pump – at least for the foreseeable future. This price disadvantage at the pump could be erased, however, if the hydrogen is put to use in a fuel-cell-powered vehicle. The greater energy efficiency of fuel cell vehicles, discussed earlier, means that only one-third to one-half as much hydrogen is needed to propel them compared with the amount of gasoline needed to propel a conventional vehicle the same distance. If the need for hydrogen fuel is cut in half due to these efficiency gains, the price of hydrogen produced by steam-reforming of natural gas would be cheaper than gasoline on a per-mile-driven basis. The same would be true of hydrogen produced from the cheapest renewable energy resources, probably biomass and wind, and even the most expensive hydrogen production pathways would be only two to three times more expensive than gasoline.

Hydrogen Vehicle Costs

Hydrogen vehicles today are all handcrafted prototype vehicles, the costs of which are not indicative of the future price of mass-produced hydrogen vehicles. For example, the hydrogen-burning "Green Car" produced by Energy Partners, Inc., cost about $180,000 to build, compared with an average price of an American-made car of about $15,000. Just the platinum used as a catalyst in a typical hydrogen fuel cell array for a vehicle built in the mid-1980s cost more than an entire conventional automobile.[12]

Hydrogen vehicle technology has not yet entered the commercial marketplace, and engineers have focused more on improving performance than on minimizing cost. The cost of hydrogen vehicles should drop sharply as the technology improves and mass production of vehicles begins. For the foreseeable future, hydrogen vehicles will be more expensive than gasoline-powered vehicles: The question is by how much.

An automobile is an interconnected system involving thousands of components. The three systems most closely related to the type of fuel used are:

- the fuel storage equipment
- the engine that burns the fuel
- the environmental controls needed to remove the pollutants generated by fuel use.

For the conventional automobile, the gasoline fuel tank costs about $200. The engine plus the transmission costs about $5,000, with the engine itself accounting for roughly two-thirds of this cost.[13] Air pollution control equipment currently adds about $500 to the cost of a car.[14] The total price of the three fuel-related vehicle systems, therefore, is about $5,700, or about one-third of the cost of an average car. (The average car sold in the United States in 1994 cost nearly $19,000.[15]) This translates to roughly $57 per horsepower for a 100-horsepower car. When measured in terms of kilowatts, the power unit commonly used to gauge electric propulsion systems, including fuel cells, the $57 per horsepower translates to about $75 per kilowatt of power (one horsepower equals 0.74 kilowatts).

Hydrogen in conventional engines versus fuel cells

Hydrogen vehicles are not likely to require pollution control equipment because they produce little or no air pollution. But both the engine and the fuel storage systems on hydrogen vehicles are likely to be more expensive than their gasoline vehicle counterparts. The cost of these systems for a fuel cell vehicle that could be produced with technology available within a decade is likely to be between $10,000 and $21,000. The overall cost for a hydrogen-burning internal combustion engine vehicle is likely to run up to $11,500.

Table 3. Cost of Vehicle Engines and Fuel Systems

Vehicle Component	Conventional Gasoline Car	Hydrogen Internal Combustion Engine Vehicle	Hydrogen Fuel Cell Vehicle
Engine	$5,000	$5,500	$9,000-$18,000
Fuel Storage	$200	$1,500-$6,000	$1,000-$3,000
Pollution Controls	$500	$0	$0
Total	$5,700	$7,000-$11,500	$10,000-$21,000

(In constant 1995 dollars)

Table 3 compares conventional gasoline fuel storage and engine costs with the costs of two types of hydrogen vehicles: those equipped with internal combustion engines and those powered by fuel cells. The table shows that redesigning a conventional gasoline-burning internal combustion engine to burn hydrogen is an inexpensive undertaking. Once large-scale production of hydrogen internal combustion engines occurs, the new engines should cost only $500 more than a conventional engine, mainly for pressure regulators and more sophisticated fuel injectors.

Replacing the gasoline fuel storage system with a hydrogen system represents the main cause of higher vehicle costs for hydrogen automobiles. Depending on the type of refueling system, a hydrogen system storing an amount of energy equivalent to conventional gasoline tanks could cost between $1,500 and $6,000 per vehicle, compared with the $200 gasoline tanks they replace. Liquefied hydrogen tanks are the cheapest storage system, while hydride and compressed hydrogen storage cylinders are the most expensive. Overall, hydrogen vehicles powered by internal combustion engines should cost between $1,300 and $5,800 more than a conventional gasoline-powered vehicle.

In fuel-cell-powered vehicles, the entire internal combustion engine and transmission are replaced with different technology. When fuel cell technology enters large-scale production for automotive use, the cost is likely to drop from its current level of between $1,500 and $2,500 per kilowatt of power down to about roughly $300 to $450 per kilowatt ($225 to $330 per horsepower). Although still many times higher than the $75 per kilowatt ($57/horsepower) cost of an internal combustion engine, the higher efficiency of fuel cell vehicles means that much smaller engines are needed to match the performance of vehicles powered by internal combustion engines. Assuming that 30- to 40-kilowatt fuel cell engines (roughly 25 to 30 horsepower) will provide comparable power to a typical car's engine, the cost of fuel cell engines for cars built within a decade will be between $9,000 and $18,000, compared with $5,000 for a conventional gasoline-burning engine. This range is well below the current cost of fuel cells, but is considerably higher than some estimates. For example, an Institute of Transportation Studies report places the range at $3,000 to $7,000.

Fuel storage costs in fuel cell vehicles, although still more expensive than gasoline tanks, will be much less than the costs of hydrogen storage in vehicles powered by internal combustion engines. Again, the improved energy efficiency of fuel cells is the reason. In terms of kilowatts stored, onboard fuel storage capacity can be cut at least in half compared with an internal combustion-engine-powered vehicle, while still providing enough energy to propel a car an equivalent distance. Nonetheless, hydrogen storage in fuel cell vehicles will cost between $1,500 and $3,000, depending on the technology employed.

As technology and production methods improve, the cost of fuel cell vehicles should drop considerably. For example, a 1994 study by General Motors concluded that proton-exchange membrane fuel cells could eventually be built for a mere $46 per kilowatt ($34/horsepower).[16] Studies cited in *Clean Fuels Report* have projected costs as low as $2,000 more than a conventional car for mass-produced fuel-cell-powered hydrogen vehicles of the future.[17]

External Cost Considerations

Besides the above-mentioned costs of purchasing and operating a hydrogen or gasoline vehicle, many other costs result from automotive use that are not directly paid by vehicle owners. These external costs, or externalities, are those beyond the costs that are internalized in payments made directly by consumers when they purchase vehicles and fuels. The combination of internalized costs plus the cost of externalities is frequently referred to as the total social cost of automotive use. Despite higher vehicle and fuel costs compared with conventional vehicles, hydrogen-powered vehicles may still be economical from a total social-cost perspective because their use entails significantly lower externality costs.

Many studies in recent years have examined the issue of externalities, and most have concluded that driving gasoline-powered automobiles entails significantly higher costs than the fuel and vehicle prices paid by owners.[18] Among the externalities most widely identified with conventional automotive use are:

- Health costs from exposure to tailpipe emissions, estimated in various studies to cost the nation $20 billion to $50 billion per year. Health costs of $320 to $17,200 have been estimated for every ton of air pollution released in 17 urban areas in the United States.[19]
- The impact of carbon dioxide emissions on global climate, estimated at $2.5 billion to $25 billion per year.
- Worldwide environmental effects of oil drilling, refinery air emissions, and oil spills, estimated at $2.5 billion to $9.0 billion per year.
- The cost of maintaining an American military presence in oil-producing regions abroad ($0.5 billion to $50 billion/year).

Studies of externality costs associated with gasoline use have reached a wide range of conclusions. Low-end estimates find that burning one gallon of gasoline results in less than $1 in externality costs. High-end estimates range from $12 to $16 for each gallon of gasoline burned in a car.[20] The findings of most studies cluster in the range of $1 to $3 as the external cost associated with burning one gallon of gasoline in an automobile. On a per-vehicle basis, one study in California has concluded that use of an electric vehicle could eliminate $17,570 in externality costs associated with the gasoline vehicle it replaced. A slightly greater externality reduction could be achieved with a fuel cell vehicle, if the hydrogen were produced from renewable resources.

Hydrogen use entails lower externalities

Since the externalities of using hydrogen as an automotive fuel would be near zero if the hydrogen is produced from renewable resources, and would still be much less than gasoline externality costs when produced from natural gas, the economics of hydrogen- versus gasoline-powered vehicles would be radically altered if some or most of the externalities associated with gasoline use were included in the price of gasoline. Even low-end estimates of $1 per gallon added to the price of gasoline would make hydrogen produced from natural gas a cheaper fuel today. On a per-vehicle basis, one study in California has concluded that use of an electric vehicle, including a hydrogen fuel cell vehicle, could eliminate $17,570 in externality costs associated with the gasoline vehicle it replaced.[21] High-end externality costs added to gasoline prices would make the total social cost of using hydrogen produced from any source a bargain. Similarly, adding externality considerations to vehicle sticker prices could eliminate the gap between the purchase price of a gasoline vehicle and a hydrogen-powered vehicle. Although the externality costs of hydrogen are clearly less than those of gasoline for most methods of hydrogen production, this is not true in all cases. For example, producing hydrogen from coal could potentially yield externalities even greater than those associated with gasoline.

Economic Credits for Hydrogen Vehicles

An alternative to adding economic penalties to the pump price of gasoline to reflect externalities is to reward the use of cleaner fuels economically. Under federal and state laws, air pollution, much of it caused by automotive exhausts, must be reduced in regions where air quality violates public health standards. These requirements are forcing governments and industries around the country to find new ways of reducing pollution, even from sources that are operating in full compliance with all applicable environmental regulations.

For example, in some areas, new or expanding industries may be required to reduce pollution at other locations equal to the amount of pollution their new facility will produce. These emission reductions are called pollution offsets. Alternatively, a government authority could set a cap on total emissions for a region, and each polluting source must find ways to reduce their pollution by a certain level as part of a strategy to reach the cap. The emissions can occur at the source itself or from another source which "trades" its pollution reductions, for a fee, for use by another party in reaching its required pollution reduction target. The amount of pollution that can be traded or sold is often referred to as emission reduction credit (ERC).

Each new pollution control measure taken under such pollution offset or trading programs comes at a cost. The science of environmental economics has developed a new discipline in recent years that attempts to quantify the cost of pollution control programs in terms of dollars spent for each pound or ton of pollution eliminated. These figures, often referred to as pollution offset costs, represent the amount of money that must be spent either directly to reduce pollution by one ton or by payment to another party for it to take action to eliminate one ton of air pollution.

Pollution offset costs vary depending on how extensive the required pollution reductions are, the cost of the individual options that are available to reduce pollution, and the competition for pollution reductions. In Los Angeles, for example, pollution offsets now being pursued by companies seeking to build new or expanded industrial facilities in the area range from \$5,000 to \$15,000, depending on the pollutant of concern.[22] New inspection and maintenance programs designed to reduce emissions from gasoline-burning vehicles are estimated to cost between \$2,600 and \$6,000 per ton of pollution eliminated.[23]

With respect to vehicle emissions, new California fuel standards cost from \$8,000 to \$12,000 per ton of hydrocarbons eliminated statewide. The federal reformulated gasoline program required for the 22 most polluted cities in the United States costs an estimated \$2,000 to \$8,000 per ton of hydrocarbons eliminated from automotive fuel burning.[24] When hydrogen vehicles are used in areas where pollution offsets are required to reduce pollution, the pollution reductions from the hydrogen vehicles compared with emissions from the gasoline vehicles they replace can be used as pollution offsets. These emission reduction credits can be substantial. One study has calculated that driving an electric vehicle would eliminate nearly

1.5 tons of the major air pollutants regulated under the Clean Air Act.[25] The value of eliminating this pollution, if sold for application as pollution offsets, could be worth up to $17,000 per vehicle.[26] As low-cost pollution reduction options are further exhausted, the relative emission reduction credit for use of hydrogen vehicles would increase, especially in highly polluted areas.

Life-cycle Cost of Hydrogen Vehicles

In recent years, several studies have attempted to quantify the total cost of owning and driving various alternative fuel automobiles. All expenditures during the projected lifetime of a vehicle are tabulated – including the initial purchase price plus finance charges, fuel costs, and a host of other operating expenses such as vehicle maintenance and insurance, and a discount factor that reflects the changing value of money over time. Dividing total expenses by total miles expected to be driven yields a life-cycle vehicle cost expressed on a cents-per-mile basis. Some life-cycle cost studies include externalities and offset credits in their analyses as well. Key parameters in life-cycle analyses that weigh heavily in the calculation are: 1) assumptions about fuel efficiency, which directly affects the amount of fuel needed, and 2) the lifespan of the vehicle, which affects the number of miles that the vehicle will be driven and vehicle depreciation expenses.

Fuel cells lower life-cycle cost of using hydrogen

A number of studies have concluded that fuel-cell-powered hydrogen vehicles, as they develop and are commercialized over a decade, will become cost-effective – entailing comparable, or more favorable, costs in comparison with gasoline vehicles.[27] In these studies, the researchers conclude that the higher cost of hydrogen would be offset in part by the externality costs of gasoline, but mainly by efficiency gains possible from using hydrogen in fuel cell engines. Similarly, the high cost of fuel cell engines compared with internal combustion engines would be largely offset by their projected 33 percent longer lives compared with internal combustion engines; the absence of combustion and moving parts in fuel cells allows them to last longer.

One 1991 study at Princeton University concluded that the life-cycle costs of operating gasoline and hydrogen vehicles would be within 5 percent of each other by the turn of the century.[28] A second Princeton paper concluded that the first generation of fuel-cell-equipped vehicles using hydrogen produced in this decade by steam reforming of natural gas will be 10 percent less costly to operate on a life-cycle basis than gasoline vehicles. The study predicted the same 10 percent cost reduction for fuel cell vehicles powered by hydrogen produced from renewable energy resources early next century.[29] A 1992 study by the Institute of Transportation Studies in Davis, California, found fuel cell vehicles powered by hydrogen produced from solar energy to be equal in cost to gasoline vehicles as soon as the price of gasoline reaches $1.43 per gallon; the breakeven cost of vehicles powered by hydrogen produced by biomass gasification is even lower – $1 per gallon of gasoline.[30] All of these studies assume significant cost reductions compared with the current cost of hydrogen vehicle technology and continued improvements in fuel cycle efficiencies. Other studies, including a 1994 report by Ford Motor Company, project less progress in achieving cost reductions needed to make hydrogen-powered vehicles cost effective. The Ford study concludes, for example, that hydrogen will not be used as an automotive fuel to any significant degree before 2020.[31]

Fuel Cycle Environmental Effects of Hydrogen

In order for hydrogen to become a viable resource in a sustainable transportation energy economy, the entire fuel cycle, from production through distribution and end use, must be environmentally sound. As the study and comparison of fuel cycle environmental effects for hydrogen and other energy resources has only recently begun in earnest, the data are sparse.[32] The analyses performed to date, however, indicate that hydrogen fuel cycles are generally far cleaner than the gasoline fuel cycle, with reductions of individual pollutants ranging from marginal amounts to nearly 100 percent. The few exceptions mainly involve land use issues, air pollution emissions associated with electricity generated using fossil fuels, and environmental concerns associated with hazardous materials used in manufacturing solar photovoltaic cells, fuel cells, and batteries.

Life-cycle air pollution for gasoline and hydrogen fuel cycles is discussed below, followed by an overview of environmental issues during fuel production and distribution, two of the three major phases in the fuel cycle. Environmental concerns during vehicle operation, the third phase, were discussed in Chapter 4.

Life-cycle Air Pollution: Renewable Resources, Fuel Cells Cleanest

Producing hydrogen from renewable resources eliminates virtually all air pollution associated with the fuel cycle. When hydrogen is produced by steam reforming of natural gas, hydrogen fuel cycles are much cleaner than conventional gasoline use, but they are not pollution-free. If the hydrogen is produced by electrolysis using electricity generated by burning fossil fuels, the environmental impact of producing hydrogen is more substantial; although emissions of most pollutants are still less than those from the gasoline fuel cycle, emissions are higher for some pollutants.

Greenhouse gas emissions, mainly carbon dioxide and methane, are the only pollutants that have been extensively studied to date with regard to fuel cycle emissions. The International Energy Agency has extensively studied greenhouse gas emissions during the fuel cycles of various transportation fuels. Its comprehensive 1993 report, *Cars and Climate Change*, concluded that a switch from gasoline to natural gas would reduce greenhouse gas emissions by 25 percent, while a switch to hydrogen produced from solar energy and burned in an internal combustion engine would reduce greenhouse emissions by 71 percent. The residual greenhouse emissions in the solar hydrogen fuel cycle studied are due to emissions generated during the production of hydrogen-powered vehicles and the energy used to transport and store hydrogen. If these uses of energy were assumed to be powered also by solar-produced hydrogen, the fuel cycle emissions would drop 100 percent compared with gasoline.[33] Another study, performed in the United States, found that greenhouse gas emissions from the solar hydrogen fuel cycle would be 82 percent below emissions from the gasoline fuel cycle if the hydrogen were burned in an internal combustion engine. The reductions would top 90 percent, however, if the hydrogen were consumed in a more efficient fuel cell engine.[34]

Producing hydrogen by steam reforming of natural gas reduces the improvement in greenhouse gas emissions to 35 percent below emissions from the gasoline fuel cycle. Producing hydrogen by electrolysis using electricity generated from burning fossil fuel yields even smaller reductions in emissions. If electrolysis is powered completely by electricity generated at coal-burning power stations, total greenhouse emissions could actually increase for transportation fuel cycles using hydrogen instead of gasoline.

Many other air pollutants besides carbon dioxide and methane are emitted during the fuel cycles of various transportation fuels. The key pollutants of concern in the gasoline fuel cycle are hydrocarbons, carbon monoxide, nitrogen oxides, and toxic substances. When hydrogen is produced from renewable resources or by steam reforming of natural gas and used in a fuel cell, emissions of all these pollutants are virtually eliminated. When burned in an internal combustion engine, some nitrogen oxide and hydrocarbon emissions are produced, but these are less than 10 percent of the emissions from a gasoline-burning engine.

Producing hydrogen with electricity generated at fossil fuel-burning power plants results in higher levels of pollution, but the reductions still offer substantial benefits over the gasoline fuel cycle. A study in the Northeastern United States found that use of battery-operated vehicles powered by electricity generated at the current mix of power plants in the region (mostly coal- or natural gas-burning, plus some nuclear and hydropower plants) would reduce nitrogen oxides by 54 percent, hydrocarbons by 90 percent, and carbon monoxide by 99.8 percent compared with gasoline vehicles.[35] A similar analysis in Los Angeles found that total fuel cycle emissions would drop about 96 percent per vehicle by switching from gasoline to electric battery-operated vehicles powered by the current mix of power plants in the region.[36] Another study placed the figure at 90 percent.[37] Applying the electricity generated in this way to produce hydrogen by electrolysis for use in electric fuel-cell-powered vehicles would achieve smaller pollution reductions, owing to the energy needed to produce hydrogen, but the improvement over gasoline vehicles would still be significant.

Fuel Production and Distribution: Hydrogen Has Less Impact than Gasoline

Gasoline: Air and water pollution

In the gasoline fuel cycle, approximately one-third of the air pollution from some key pollutants (notably nitrogen oxides) occurs during oil production, refining, and distribution.[38] Most water pollution from the gasoline fuel cycle, including oil tanker spills and leaking from underground storage tanks, also occurs during these early stages.

The fuel production component of the gasoline and diesel fuel cycles consists of the drilling and extraction of crude oil from the ground and its conversion into gasoline and diesel fuels at oil refineries. The sources of emissions from the production phase are pipeline leaks; oil evaporation during extraction from the ground; and emissions from the equipment used to drill oil wells and to operate them once oil is flowing. As oil wells age, secondary and tertiary recovery technologies are often employed. These involve the injection of steam or other substances into the ground to increase oil output, thereby adding more greenhouse emissions in the process. Manufacturing the equipment used in these stages – from oil well casings to refinery cracking towers – provides an additional, indirect source of greenhouse gas discharges.

Other major environmental impacts from the production of oil-derived fuels concern the tremendous emissions of other air pollutants, including emissions of highly toxic chemicals, produced by oil refineries. Oil wells themselves mar the land surface and contaminate underground water reservoirs by rupturing water aquifers and permitting the entrance of oil and other chemicals. Oil well operations release hydrocarbon, nitrogen oxide, and other non-greenhouse gas emissions.

The distribution component of the gasoline fuel cycle releases further air pollution. Distribution consists of the transportation of oil from the site of production to the oil refinery and the shipment of the refined fuels to automotive refueling stations. Emissions of air pollutants during the distribution of oil-derived fuels result primarily from the burning of fuel by tanker ships, trains, and trucks. The heaters and compressors that keep oil-products liquefied and moving through pipelines also produce emissions.

Apart from air pollution, water pollution is clearly the most significant environmental effect of oil distribution. The destruction of the marine environment through oil tanker spills and the contamination of groundwater supplies by the hundreds of thousands of leaking underground fuel storage tanks are two of the world's most severe water pollution problems. Both are the result of the distribution of oil and its refined products to energy markets, primarily automobiles.

Hydrogen production: Land use and toxicity concerns

The production and distribution components of the hydrogen fuel cycle are superior environmentally to the current technologies used to extract and refine oil. How superior depends on what primary energy resource is used to produce the hydrogen. The most dramatic environmental benefits of hydrogen are evident when it is produced from renewable energy resources, especially from the direct use of solar energy. Emissions of most air pollutants drop to near zero. Small emissions of greenhouse gases occur during the manufacturing of the equipment used in renewable energy technologies and in the operation of these systems, but this is also true of the gasoline fuel cycle.

The biggest environmental issue raised by producing hydrogen from renewable resources is not air pollution but land-use impact. The largest impact on land use is obviously posed by hydropower projects which involve the damming of rivers and the creation of huge artificial lakes, interfering with natural aquatic and river floodplain ecosystems. The long-term environmental devastation caused by dams has led to strong opposition to expanded use of hydropower technology despite its environmental benefits, which include the absence of air pollution and the renewable nature of the resource.

Among the other renewable energy resources, biomass has the largest land use requirements.[39] Nearly 60 million acres of land, close to 100,000 square miles, would be needed to produce enough biomass resources to supply hydrogen to the American automotive fleet in the year 2010 – if every vehicle was equipped with a hydrogen fuel-cell engine. As large as this figure is, however, it represents only 70 percent of the cropland that lay dormant in 1990. Solar thermal power plants also have large land use requirements, about one square mile for each 100 megawatts of generating capacity, but this is only half the space needed by biomass-based systems. Wind power systems occupy about the same land area as solar thermal systems, but they entail an additional problem: the danger posed to birds that inadvertently fly into the turbine blades.[40] However, windmills kill many fewer birds than do overhead electrical transmission lines, or, for that matter, domesticated cats.

Solar photovoltaic technology demands the least space of the major renewable technologies. Even so, generating enough hydrogen to fuel American cars in the year 2010 would require 4,000 square miles of solar panels. Unlike hydropower systems, these other renewable technologies temporarily occupy, but do not devastate, the land on which they are located. The land surface can return to its natural state when the renewable energy systems are dismantled.

The use of highly toxic materials in the manufacturing of some renewable energy technologies presents another potentially significant environmental consequence. This is especially true for solar photovoltaic and electrolysis systems, which often contain toxic metals and solvents. Manufacturing procedures for these technologies are similar to those used in the semiconductor or computer industries. Steps must be taken to ensure that the workplace environment remains safe and the usage of the equipment does not lead to adverse environmental effects.

The hydrogen distribution system is clearly an improvement over the gasoline system. Tankers filled with liquefied hydrogen are the most likely form of transportation across oceans. In the event of a crash, the fuel would vaporize and disperse upward into the atmosphere. Although emissions during this type of transportation are minimal, the process of liquefying hydrogen for more energy-dense distribution increases its environmental impact. As liquefaction can consume up to 40 percent of the energy in the hydrogen, more fuel must be used to power the liquefaction process.

Pipelines are the most likely transportation method to move large amounts of hydrogen across land. The land-use impacts of hydrogen pipelines are similar to those of oil pipelines, although ruptures of hydrogen pipelines would create less environmental damage than oil pipeline breaks because of the rapid dispersion of the gas. The same result would occur with truck or barge tankers filled with hydrogen in a liquefied or compressed state. Bulk storage options for hydrogen at the end of the distribution line include unlined salt caverns, depleted aquifers, or natural gas reservoirs.

Hydrogen Safety

The reputation of hydrogen as an unsafe transportation fuel has been established quite firmly, but wrongly, in the eyes of the public by the vivid images captured on film of the ferociously burning Hindenburg dirigible in 1937 and the self-destructing Challenger space shuttle in 1986. Although burning hydrogen certainly featured in both tragedies, it was not the cause of death in either one.

In reality, throughout decades of use as a major chemical feedstock and as a major transportation fuel, hydrogen has compiled an excellent safety record, as described below. This record does not automatically mean, however, that hydrogen usage does not raise very important safety issues or that hydrogen production and use can be scaled up to supply massive demand for energy in transportation without additional safety precautions. Technological solutions to the safety challenges posed by hydrogen are either available or are under development, but the lack of standard safety codes has thus far inhibited hydrogen's commercialization as a transportation fuel.

The Poor Safety Record of Gasoline

While hydrogen's safety record has been very good, the same cannot be said of the safety record of conventional gasoline-powered automobiles and trucks, which present a safety risk to which most people have become inured. Each year, more than 15,000 automobiles are destroyed by engine fires. Another 2,500 fires occur at gasoline refueling stations. Some 500 deaths in auto accidents are attributed solely to burns from ignited fuel leaks from gasoline fuel tanks; an additional 7,000 people are injured. The accident rate for the ubiquitous gasoline tankers thundering day and night along public roadways is similarly disturbing. Statistics collected by the Department of Labor show that, per unit of energy carried, gasoline tanker transportation is 1,000 times more dangerous than natural gas pipeline transportation.[41]

If the early automotive innovators were asked 100 years ago to make a decision based solely on safety concerns about which fuel should be used to power automobiles in the future, they almost certainly would not have selected gasoline. Availability, convenience, and cost favored gasoline at the turn of the century, and these conditions continue today, creating the great barriers to alternative transportation fuel use that now exist. Gasoline has been and continues to be a dangerous transportation fuel. It will not be easy to make hydrogen the safest of all possible fuels of the future, but if the target is the current record of gasoline, a safe transportation system based on hydrogen is achievable.

The Hydrogen Safety Myth

Because it is lighter than air, huge amounts of hydrogen were used in the Hindenburg's 16 air bags to keep the dirigible, or blimp, aloft during transatlantic journeys from Europe to the United States.[42] The hydrogen was used solely for buoyancy; the airship was actually propelled by diesel-powered engines that turned 20-foot wooden propellers during its 3 to 5 day trips across the ocean. The fateful trip that ended with the ship's destruction was its tenth journey. There were 97 people on board, in the passenger gondola suspended beneath the inflated bags of hydrogen, when the Hindenburg attempted to dock at Lakehurst, New Jersey, in midsummer 1937. As the docking equipment was being attached to the floating airship, a fire broke out that quickly ignited the hydrogen. A ball of flame rose 500 feet into the evening sky. In a mere 26 seconds, the entire ship was reduced to a smoldering, burned-out wreckage of twisted metal. In all, 35 people died. The cause of the fire remains unknown. It could have been the result of a lightning strike or smaller electrical discharge from the thunderstorms in the area. Other possible causes include an ignition of spilled diesel fuel, accidentally ignited cabin furniture, an onboard electrical fire, or sabotage.

Although widely and understandably viewed with horror, the Hindenburg accident reveals more about the safety of hydrogen than it does about its hazards. Most importantly, nearly two-thirds of the people on board, 62 of the 97 travelers, survived. They survived by remaining on the ship. Most of the casualties, 27 of the 35 deaths, came as terrified people jumped from the ship while it was still aloft. The remaining 8 deaths resulted from burns received by exposure to burning diesel fuel or cabin furnishings in the gondola. The fact is, the huge hydrogen fireball itself killed no one. Being a light gas, the flaming hydrogen rapidly dispersed upward, away from the people. Moreover, unlike gasoline and diesel, burning hydrogen radiates very little heat. Hydrogen flames are very hot, but the heat is imperceptible to those standing even a short distance away. The 62 survivors were the passengers who did not panic and who were not doused with burning diesel fuel. They survived by waiting until the gondola sank to the ground, allowing them to run away safely. A recent reexamination of the incident noted that if the energy content of the hydrogen aboard the Hindenburg been in the form of gasoline, the fire would have killed not only all 97 passengers, but the 200-person ground crew and observers on hand as well.[43]

Hydrogen is a fuel and, like all fuels, it must be handled with care. Filling a dirigible's huge air bags with any fuel to achieve buoyancy is lunacy, equivalent to loading the back seat of your car with open tubs of gasoline and lighting a cigarette while sitting in the front seat. The risk was known to the German owners of the Hindenburg. Before 1937, dirigibles were filled with helium, another gas that is also much lighter than air, but chemically inert. Helium will not react with any chemicals and cannot be ignited.

In the 1930s, the United States was the world's only producer of helium gas. In the tense political climate of the pre-World War II years, the

United States had instituted an economic embargo of Germany and stopped all helium exports. Faced with the alternative of grounding its dirigible fleet, Germany chose – wrongly, in retrospect – to replace inert helium with highly flammable hydrogen. Had the Hindenburg's owners continued to respect hydrogen and to treat it like a fuel, the disaster never would have occurred.

The other major transportation application of hydrogen to date has been as the principal fuel for the space program. During the first few minutes of takeoff, the Space Shuttle burns liquefied hydrogen at a rate of 15,000 gallons per minute. At lift-off, the Shuttle's hydrogen tanks are filled with 385,000 gallons of fuel. The Challenger disaster, which claimed the lives of the entire six-member crew in 1986, had nothing to do with the hydrogen fuel system, however. The infamous O ring seal that ruptured, causing the destruction of the spaceship, was part of the solid-fuel booster rocket, not part of the hydrogen-fueled main engine system. Similarly, the fire that claimed the lives of three astronauts during a 1978 training exercise in Texas (which, until the Challenger disaster, was the largest mishap in the history of the space program), was caused by a malfunction in the electrical system serving the cockpit. The fire was fanned by the pure oxygen atmosphere in the training capsule, engulfing the astronauts within seconds as it burned through the cabin's combustible materials. Hydrogen was not involved.

The Hydrogen Safety Reality

Because hydrogen is a powerful fuel, containing by far more energy per pound than any of the fossil fuels, the potential for serious accidents with hydrogen is always present, as it is for all fuels. Expanding the use of hydrogen requires analyzing safety issues step by step throughout the hydrogen fuel cycle and developing standards to ensure that the fuel is always handled in a manner that reduces the risk of harm.

Some chemical and physical characteristics make hydrogen a more hazardous fuel relative to gasoline and other alternative transportation fuels, while others make it safer.[44] Some traits work as either safety advantages or disadvantages, depending on the environment in which it is used.

Safety disadvantages of hydrogen

Under many conditions, hydrogen is more likely to ignite accidentally due to a combination of four physical characteristics of the fuel:

- Low ignition temperature
- Wide flammability range
- Propensity to diffuse rapidly
- Fast flame speed.

The low ignition temperature and the wide flammability range of hydrogen are probably the two most troublesome traits with regard to safety. Hydrogen requires less than one-tenth the energy needed to ignite either gasoline or natural gas – only 19 Btu. Moreover, hydrogen burns when mixed in a wide range of air and fuel concentrations, from 4 to 75 percent hydrogen. Flammability concentrations for gasoline and natural gas are much more narrow, ranging from 1 to 7.5 percent and 5.3 to 15 percent, respectively. Mixtures of air and hydrogen that contain less than 4 percent hydrogen are too dilute, or "lean," to burn. This circumstance would be encountered if hydrogen leaked into the open air on a windy day; dispersion in the air would rapidly dilute the fuel to below a 4 percent concentration. At the other end, mixtures with more than 75 percent hydrogen are too "rich" with fuel to burn. This would include hydrogen contained in compressed gas storage containers or hydrogen leaked into an enclosed space.

Most ratios of air and hydrogen, however, are ignitable. Within an internal combustion engine, the wide flammability range is an advantage because it permits engines to be designed to achieve maximum performance even under very lean operating conditions, where fuel efficiency is at its greatest. The combination of wide flammability and low ignition temperature is problematic, however, because leaks and unintended discharges of hydrogen into the atmosphere could easily ignite.

Adding to this safety concern is a third characteristic of hydrogen, its high diffusion coefficient, or propensity to diffuse rapidly. A high diffusion coefficient means that hydrogen tends to leak through even very tiny openings, such as at pipe joints and storage tank seals. Hydrogen has been shown in experimental conditions to leak 4 to 6 times faster than natural gas through pipelines, although the leakage rates in both cases were found to be negligible. Because hydrogen can deteriorate certain metals through embrittlement, there is some concern that underground hydrogen pipelines could rupture, creating explosions.[45]

Hydrogen's fast flame speed presents a fourth safety concern. Once ignited, a flame front of burning hydrogen moves much faster than either gasoline or natural gas. The flame velocity of hydrogen is 108 inches per second, versus approximately 15 inches per second for natural gas and gasoline. In a confined space, the fast flame speed of hydrogen can cause the mixture to detonate, or explode, more readily than in comparable circumstances where gasoline or natural gas is the burning fuel.

When combined, these four physical characteristics present a difficult challenge to maintaining the safety of hydrogen systems. Hydrogen is more likely to leak than most fuels because of its high diffusion coefficient; its accidental presence is more likely to be combustible because of its wide flammability limits; a flammable mixture of hydrogen and air is more likely to catch fire because of its low ignition energy; and, once ignited, fires spread very rapidly because of hydrogen's fast flame speed – and the fuel may even explode.

Further problems are created because hydrogen, like natural gas, is both colorless and odorless. Moreover, hydrogen burns with an almost invisible flame. These three characteristics make the presence of hydrogen, even when burning, harder to detect, avoid, or control than other fuels.

Safety advantages of hydrogen

Positive safety aspects of hydrogen derive both from its inherent characteristics and from the requirements of the technologies to use it:

- Low density of hydrogen gas
- Low heat transmission of burning hydrogen
- Stronger, more secure storage and distribution systems
- Fuel cells' lack of combustion.

Whereas hydrogen's low density, and thus propensity to diffuse rapidly, is problematic in underground pipeline transportation, it offers a safety advantage in the event of a hydrogen vehicle accident. The lightness of hydrogen – only 7 percent as dense as air – partially offsets the risk of accidental ignition. Once released, hydrogen will quickly disperse upward, moving away from the point of discharge even faster than natural gas. Gasoline, on the other hand, forms liquid puddles which, once ignited, burn furiously near the point of discharge. Moreover, as mentioned above, burning hydrogen transmits only half the energy of combustion to the surrounding air in the form of heat as does burning of gasoline – about 20 percent versus 40 percent, respectively.

Other advantages of hydrogen are derived from the characteristics of the technologies that are being developed to handle hydrogen. Pipeline embrittlement, for example, must be prevented in order for pipelines to be economically viable, regardless of the safety risks that hydrogen leakage from deteriorating pipelines may present. Similarly, all hydrogen storage technologies are designed to be impervious to leaks and ruptures to a degree far exceeding those for thin-walled tanks used to store conventional liquid fuels. Tanks for compressed hydrogen, for example, must be thick and strong to withstand the high pressures; liquefied hydrogen tanks must be very heavily insulated and airtight to prevent warming of the fuel; and hydride and carbon adsorbent systems bind the hydrogen directly to solids, preventing fuel leakage even if the systems are ruptured in a crash.

Fuel cell engines offer safety advantages for hydrogen use compared with use in internal combustion engines because no fuel-burning takes place and no high engine temperatures are produced – although acids from some types of fuel cells that use liquids might leak out in the event of a vehicle crash, as they might with batteries. The best single approach to reducing hydrogen safety risks is through the development of fuel cell technology for its use. The rewards of this approach multiply into the production of better performing, more efficient, and safer hydrogen systems.

Technical solutions to safety challenges Specific technologies are available or under development to address each of the safety concerns raised by hydrogen use. These include:

- Strong odorant additives
- Additives for visibility and to reduce explosion risk
- Alarm detection systems

- Improved ventilation and containment systems
- Hydrogen-capturing canisters
- Embrittlement inhibitors
- New pipeline materials.

Whether these systems can adequately reduce the risks of hydrogen use to overcome public concern is one of the many unanswered questions facing hydrogen.

Strong odorants can be added in very small quantities to hydrogen, as they are to natural gas, to enable detection by smell. This does not create problems if hydrogen is ultimately burned – the odorants also burn with virtually no resulting air pollution – but odorants can contaminate hydride or carbon adsorbent fuel storage systems, poison fuel cell engines, or clog liquefied hydrogen systems. Additives can make hydrogen flames visible or less susceptible to detonation, but they may similarly interfere with fuel storage systems and fuel cell engines. More research is needed to resolve this issue.

Hydrogen detection systems, which sound an alarm when they detect the gas in concentrations well below the 4 percent flammability limit, offer another safety precaution. They provide the best insurance, at relatively little cost, against accidents from leaking hydrogen as long as they are working properly. Improved ventilation and hydrogen containment systems are also being developed to handle accidental releases of hydrogen. A prototype hydrogen-powered car built by BMW, for example, is equipped with a crash detector similar to the system which now activates automotive safety air bags. When a crash in the hydrogen car is detected, the windows, doors, and trunk cover of the car automatically open to promote rapid dispersion of any hydrogen that may be released as a result of the crash. Similarly, hydrogen-capturing canisters are being tested that cleanse the air of hydrogen with miniature fuel cells that convert the hydrogen into water. These systems can be placed along the ceilings of garages or near fuel storage tanks, where they can capture hydrogen that may, for example, be vented as boil-off from a liquefied hydrogen storage system.

Embrittlement inhibitors capable of preventing pipeline deterioration are under development. Similarly, new pipeline materials – including joint fittings – that both reduce hydrogen leaks through diffusion and prevent embrittlement, are being studied.

Technology now exists to ensure that handling hydrogen can be accomplished at least as safely as handling other transportation fuels. This conclusion has been reached in a number of hydrogen safety studies at the Clean Energy Research Institute at the University of Miami.[46] Moreover, a 1993 safety analysis performed by the German Federal government, entitled "Risks of Increased Use of Hydrogen," concluded that "the technical risks in all parts of a hydrogen energy system, from production to utilization, are in principle, regarded as controllable."[47] Another German safety analysis, focusing on risks from spills involving liquefied hydrogen, came to the same conclusion.[48]

A key obstacle to the commercialization of hydrogen is the absence of widely recognized safety codes regulating its production, distribution, and use as a transportation fuel. Safety standards now exist for particular aspects of hydrogen handling, including cylinder and pipeline standards, but these regulations do not cover all aspects of the hydrogen fuel cycle and they were set with small-volume industrial applications in mind.[49] To address this need, the Swiss-based International Standards Organization (ISO) established Technical Committee 197 in the late 1980s to develop a comprehensive set of hydrogen regulations. The work of this committee, which includes representatives from the United States and is now headed by representatives from Canada, is continuing. In a separate action, in January 1995, a hydrogen safety workshop convened in Washington, DC. Its 40 participants concluded that a standard-setting process parallel to the ISO undertaking should be initiated in the United States.[50]

Notes

1. The principal source of information for the energy efficiencies discussed in this section came from "Appendix A: Energy Pathways," contained in the *Hydrogen Program Plan: FY 1993 – FY 1997*, United States Department of Energy (Washington, DC, June 1992). Other sources include: J. Badin and S. Tagore, "Energy Pathway Analysis – A Hydrogen Fuel Cycle Framework for System Studies," *Hydrogen Energy Progress X: Proceedings of the 10th World Hydrogen Energy Conference* (Cocoa Beach, FL, June 20-24, 1994); D. Illman, "Automakers Move Toward New Generation of 'Greener' Vehicles," *Chemical and Engineering News*, August 1, 1994; D. McCosh, "Emerging Technologies for the Supercar," *Popular Science*, June 1994; and "Piston-Engine Hydrogen Hybrid Could be as Efficient as a Fuel Cell," *Clean Fuels Report*, April 1994.
2. D. Sperling, *Future Drive: Electric Vehicles and Sustainable Transportation* (Island Press, Washington, DC, 1995), p. 47.
3. *Ibid.*
4. R. Riley, *Alternative Cars in the 21st Century* (Society of Automotive Engineers, Warrendale, PA, 1994), p. 96.
5. J. Smith. "The Hydrogen Hybrid Option" (Lawrence Livermore Laboratory, Livermore, CA, October 15, 1993).
6. *Ibid.*
7. Stacy Davis, *Transportation Energy Data Book: Edition 14* (Oak Ridge National Laboratory, Oak Ridge, TN, 1994), p. B-2.
8. United States Energy Information Administration, *Annual Energy Review 1993* (Washington, DC, 1994), p. 181.
9. D. Gushee, *Impact of Highway Fuel Taxes on Alternative Fuel Vehicle Economics* (Library of Congress, Washington, DC, March 16, 1994), p. 2.
10. Hydrogen fuel and vehicle costs appearing in Tables 2 and 3 were obtained from a number of studies, the most comprehensive of which include the following: M. DeLuchi, *Hydrogen Fuel-Cell Vehicles* (Institute for Transportation Studies, Davis, CA, 1992); J. Ogden and M. DeLuchi, "Renewable Hydrogen Transportation Systems," *Hydrogen Energy Progress IX: Proceedings of the 9th World Hydrogen Energy Conference* (Paris, France, June 22-25, 1992); D. Swan and D. Smith, *Hydrogen-Fueled Vehicles: Technology Assessment Report* (California Energy Commission, Sacramento, CA, 1991); and M. DeLuchi, "Hydrogen Vehicles: An Evaluation of Fuel Storage, Performance, Safety, Environmental Impacts, and Cost," *International Journal of Hydrogen Energy*, Vol. 14, No. 2, February 1989.

11. James S. Cannon, "Reformulated Gasoline: Cleaner Air on the Road to Nowhere" (INFORM, Inc., New York, 1994).

12. C. Borroni-Bird, "Fuel Cells on a Future Automotive Powertrain," presented at *Proceedings of the SAE Fuel Cells for Transportation TOPTEC Conference* (Santa Fe, NM, March 29, 1995).

13. D. Swan, "Fuel Cell Powered Electric Vehicles," SAE paper #891724 (Society of Automotive Engineers, Warrendale, PA, 1989); Swan and Smith, *Hydrogen-Fueled Vehicles: Technology Assessment Report, op. cit.,* pp. 29-30; and J. Appleby, "Fuel Cells and Hydrogen Fuel," *International Journal of Hydrogen Energy*, Vol. 19, No. 2, February 1994.

14. "Current Emission Control Costs Calculated," *Clean Fuels Report*, April 1993.

15. Projection from American Automobile Manufacturers Association, *Facts and Figures 1993* (Detroit, MI, 1994).

16. Allison Gas Turbine Division of General Motors, *Research and Development of Proton-Exchange Membrane (PEM) Fuel Cell System for Transportation Applications*, prepared for the United States Department of Energy, Office of Transportation Technologies, November 1993; and "GM Study Shows Non-Polluting Fuel Cell Engine Can be Made at No Extra Cost," press release, Fuel Cells 2000, 1993.

17. "Solar Hydrogen Seen as Sustainable Energy Option," *Clean Fuels Report*, February 1994.

18. Among the more comprehensive externality studies used in writing this section are the following: Congress of the United States, Office of Technology Assessment, *Saving Energy in U.S. Transportation* (Washington, DC, 1994); J. MacKenzie *et al., What It Really Costs to Drive* (World Resources Institute, Washington, DC, 1992); James S. Cannon, *The Health Costs of Air Pollution* (American Lung Association, New York, 1990); H. Hubbard, "The Real Cost of Energy," *Scientific American*, Vol. 264, No. 4, April 1991; and Danish Ministry of Energy, *Monetary Evaluation of Transport Environmental Impact – The Case of Air Pollution* (Lyngby, Denmark, 1991).

19. "Monetary Values Calculated for Air Pollutant Reductions in 17 Cities," *Clean Fuels Report*, April 1995.

20. R. Riley, *Alternative Cars in the 21st Century, op. cit.,* pp. 24-25.

21. R. Hwang *et al., Driving Out Pollution: The Benefits of Electric Vehicles* (Union of Concerned Scientists, Berkeley, CA, 1994), p. 24.

22. B. Beebe, "Transportation Credits: Are They Real," *Proceedings of the 4th Annual Meeting of the National Hydrogen Association* (Washington, DC, March 25-26, 1993); and "Emission Reduction Credits for Mobile Sources are Governed by Complex Rules," *Clean Fuels Report*, September 1994.

23. "Resources for the Future Questions Cost-Effectiveness of Alternative Fuels," *Clean Fuels Report,* June 1995.

24. "Committee for Economic Development Opposes Extension of RFG and Alternative Fuel Mandates," *Clean Fuels Report*, June 1993; and *An Evaluation of the City of New York's Alternative Fuel Vehicle Program for Fiscal Year 1994* (New York City Alternative Fuels Task Force Report, September 15, 1994), p. 64.

25. R. Hwang *et al., Driving Out Pollution..., op. cit.,* p. 3.

26. "Calstart Official Praises California's Tough Clean Air Rules at CARB Hearing," *Calstart News,* May 12, 1994, cited in *Alternative Energy Network Online Today.*

27. See citations in note number 10 plus: "Real Costs of Hydrogen Said to Be Less than Gasoline," *Clean Fuels Report*, February 1991; "Prospects for Solar Hydrogen as a Transportation Fuel Reviewed," *Clean Fuels Report*, September 1993; "Solar Hydrogen Energy to Be the Choice of the Future," *Clean Fuels Report*, February 1992; "Fuel Cell Vehicles Could Allow Hydrogen to Compete with Gasoline," *Clean Fuels Report*, February 1994; and "Argonne Compares Studies in Cost Effectiveness of Alternative Fuels," *Clean Fuels Report,* June 1995.

28. "Princeton Study Shows Competitive Life Cycle Costs for Fuel Cell Vehicles," *Clean Fuels Report*, November 1991.

29. "Hydrogen from Renewable Energy Seen as Attractive Early Next Century," *Clean Fuels Report*, September 1992.

30. "Hydrogen Fuel-Cell Vehicles Look Good in New Analysis," *Clean Fuels Report*, February 1993.

31. "Use of Hydrogen as an Automotive Fuel Updated," *Clean Fuels Report*, June 1994.

32. The primary sources for this section were DeLuchi, *Hydrogen Fuel Cell Vehicles, op. cit.;* Hwang *et al., Driving Out Pollution..., op. cit.;* and M. DeLuchi and J. Ogden, *Solar-Hydrogen Fuel-Cell Vehicles* (University of California Transportation Center, Berkeley, CA, 1993)

33. International Energy Agency, *Cars and Climate Change* (Paris, France. 1993), p. 96.

34. M. DeLuchi, *Emissions of Greenhouse Gases from the Use of Transportation Fuels and Electricity* (Argonne National Laboratory Center for Transportation Research, Argonne, IL, 1991), p. 61.

35. "Union of Concerned Scientists Analyst Heralds EVs as Cleaner Air Solution for Northeast," *Boston Globe*, September 20, 1994, reported by *Alternative Energy Network Online Today.*

36. "Analysis by California Air Resources Board Affirms Benefits of Zero Emission Vehicles," *Clean Fuels Report*, April 1994.

37. "EDF/NRDC Study Reports EVs Really Are Cleaner," UPI Release, June 8, 1994, reported by *Alternative Energy Network Online Today.*

38. Energy International Inc., *Light Duty Vehicle Full Fuel Cycle Emissions Analysis* (Gas Research Institute, Chicago, IL, April 1994), p. 9.

39. DeLuchi, *Solar-Hydrogen Fuel-Cell Vehicles, op. cit.*, p.72; and J. Ogden and J. Nitsch, "Solar Hydrogen," in *Renewable Energy: Sources for Fuel and Electricity* (Island Press, Washington, DC, 1993).

40. P. Asmus, "Hot Air, Hot Tempers, and Cold Cash: Clashes of Ethics and Clashes of Interests in the Controversy Over Wind Power," *The Amicus Journal*, Fall 1994.

41. James S. Cannon. *Paving the Way to Natural Gas Vehicles* (INFORM, Inc., New York, 1993), p. 10.

42. The Hindenburg disaster is discussed in a number of hydrogen energy studies including: J. O'M Bockris *et al.; Solar Hydrogen Energy: The Power to Save the Earth* (MacDonald & Company, London, 1991); and H. Braun, *The Phoenix Project: An Energy Transition to Renewable Resources* (Research Analysts, Phoenix, Arizona, 1990).

43. "Professional Engineer Reviews the Hindenburg Incident," *Clean Fuels Report*, June 1993.

44. The major information source for the section on hydrogen safety was: J. Hansel *et al.,* "Safety Considerations in the Design of Hydrogen-Powered Vehicles," *International Journal of Hydrogen Energy*, Vol. 18, No. 9, September 1993. Safety issues are also discussed in nearly all of the comprehensive hydrogen studies listed in the bibliography.

45. M. Swain and R. Adt, "Safety of Hydrogen Systems," in *Solar Hydrogen Energy Systems*, prepared by the Clean Energy Research Institute, University of Miami, October 1989.

46. "Hydrogen Safety Concerns Alleviated," *Clean Fuels Report*, September 1991; and "Hydrogen Pipeline Leaks Would Be Less Hazardous than Natural Gas," *Clean Fuels Report*, February 1994.

47. "Widespread Hydrogen Use Likely to be Low-Risk," *The Hydrogen Letter*, June 1993.

48. "Spilling LH_2 Seems Unproblematical, Test Shows," *The Hydrogen Letter*, September 1994.

49. Air Products and Chemicals, Inc., "Safetygram-4: Gaseous Hydrogen," not dated.

50. "Hydrogen as Energy Needs Specific Safety Rules," *Hydrogen & Fuel Cell Letter*, February 1995.

Chapter 8

United States Programs to Promote Hydrogen Use in the Transportation Sector

Since 1990, interest in hydrogen as a fuel has grown dramatically in the United States. The federal research and development program for hydrogen technologies, although still minuscule in comparison with other energy development efforts, has increased tenfold since the beginning of the decade. In 1995, the federal budget for the National Hydrogen Program totalled $10 million. Hydrogen-related research being performed under a number of other federal transportation programs raises total federal expenditures for hydrogen to several times this amount. Promising state and local government hydrogen vehicle development programs are also underway.

The most exciting arena for hydrogen development is probably the private sector, where a number of entrepreneurs have formed small companies to develop and market truly innovative automotive technology using hydrogen in fuel cell engines. Taken together, public- and private-sector efforts in the United States are among the most ambitious in the world, placing the country behind only Japan and Germany in terms of total annual investment in hydrogen.

The current interest in hydrogen energy in the United States builds on work performed during two previous decades: the 1960s, when the space program developed hydrogen as a key fuel to propel spacecraft into orbit and for use in fuel cells to provide electricity while in space, and the 1970s, when funding for research into various nonconventional alternative energy sources, including hydrogen, increased in the aftermath of the 1973

OPEC oil embargo. Both efforts languished in the 1980s as severe federal government cutbacks in the space and alternative energy programs slashed public- and private-sector spending on hydrogen.

Environmental and energy security concerns, especially in transportation, have spurred renewed interest in hydrogen in the United States. Hydrogen-powered vehicles, including those using hydrogen internal combustion engines, fuel cells, and hybrid engine configurations, are now undergoing field testing; a number of new vehicles are being designed for demonstration during the next few years.

More Investment, Competition, and Political Leadership Needed

Despite the encouraging upswing in the United States, the total effort is still far too small to catalyze a move toward a sustainable transportation system based on hydrogen. The United States spent 5,000 times more money in just a few months during the 1991 Gulf War that helped secure foreign oil supplies than it spends annually on the National Hydrogen Program, which could end the nation's dependence on oil.

A greater commitment of federal dollars – already the largest source of funds sustaining hydrogen research and development – would help ignite private-sector involvement in hydrogen technology. But a redirection of funding priorities to stimulate a competitive environment in the automotive industry, and political leadership to make hydrogen research a national priority, are also vitally needed.

To allow entrepreneurs to enter the field traditionally reserved for the major automotive manufacturers, the federal government could establish performance goals for advanced vehicles and fund a large number of private-sector programs aimed at achieving them. Demonstration projects funded, but not "micro-managed," by the federal government are currently the best way to show that hydrogen vehicles can be built and operated in real-world driving conditions. Demonstrations show policymakers and the public that hydrogen vehicles provide a viable alternative to continued reliance on conventional fuels and vehicle technology.

Currently, the government focus is on partnerships for new vehicle development that are directed and largely performed by the major automotive manufacturers. The technical expertise of these companies is established. However, asking the manufacturers of conventional automobiles to lead the effort to invent advanced vehicle technology seems as improbable as Henry Ford choosing the owners of horse stables to head his automotive research department a century ago, with the rationale that horse owners knew the most about current transportation practices.

Of all the roles the federal government can fulfill, political leadership is perhaps the most important. The automobile is one of the most pervasive devices defining American life today – affecting environmental quality, energy use patterns, and economic development. To embrace the multidimensional nature of the issues involved, leadership over a fundamental change in the automobile must come from the highest level. Moving beyond specialized interests, the White House could endorse and promote a national commitment to spur the United States toward a sustainable transportation system and to prepare it to compete with other nations in the automotive markets of the 21st century.

Federal Hydrogen Vehicle Program Activities

Prior to 1990, the Department of Energy sponsored hydrogen research and development activities that totalled only a few million dollars per year. By 1994, hydrogen programs had grown about tenfold, to more than $30 million per year. Although this increase is dramatic, total expenditures remain only a little more than one-tenth of 1 percent of the Department of Energy's $18 billion annual budget – about two-thirds of which is allocated for nuclear weapons activities. Even today, DOE spends 90 times more money on petroleum research than it spends on hydrogen.[1] Nonetheless, several transportation initiatives established or expanded in 1993 make the prospects for a much expanded hydrogen vehicle effort more promising than at any previous time.

Figure 1. Major Federal Hydrogen Research & Development Activities

National Hydrogen Program *Administered by: Department of Energy Office of Utility Technologies*

Activity:	**Hydrogen Storage**
Participants:	Syracuse University National Renewable Energy Lab University of Hawaii Florida Solar Energy Center Lawrence Livermore and Sandia National Labs H Power Corp. Energy Conversion Devices Arthur D. Little, Inc.
Activity:	**Hydrogen Fuel Production**
Participants:	University of Miami Oak Ridge and Lawrence Livermore National Labs National Renewable Energy Lab
Activity:	**Development and demonstrations of hydrogen in vehicles and utility and industrial applications**
Participants:	Los Alamos, Lawrence Livermore, and Sandia National Labs
Activity:	**Analyses of hydrogen fuel cycle energy efficiency, safety, and transportation and fueling infrastructure**
Participants:	Princeton University National Hydrogen Association University of Miami Florida Solar Energy Center Energetics, Inc.

National Fuel Cells in Transportation Program *Administered by: Department of Energy Office of Transportation Technologies*

Activity:	**Heavy-duty hydrogen fuel cell engine development**
Participants:	Department of Transportation South Coast Air Quality Management District H Power Corp. Booz, Allen & Hamilton Transportation Manufacturing Corp. Fuji Electric Soleq Corp. Georgetown University
Activity:	**Fuel-cell-powered passenger vehicle development**
Participants:	Allison Gas Turbine Division of General Motors Los Alamos National Lab Ballard Power Systems Dow Chemical
Activity:	**Onboard steam reforming of methanol to hydrogen**
Participant:	Arthur D. Little, Inc.
Activity:	**Onboard hydrogen storage**
Participant:	Arthur D. Little, Inc.

The primary federal effort to develop hydrogen as a domestic energy resource is the National Hydrogen Program, while the major initiative to develop hydrogen in fuel cells takes place under the Fuel Cells in Transportation Program. The Advanced Utility Concepts Division within the Office of Utility Technologies in Washington, DC, has responsibility for developing and implementing the National Hydrogen Program. The National Renewable Energy Laboratory in Golden, Colorado, serves as the technical manager of the program and coordinates the activities of the contractors.

A separate domain within the Department of Energy – the Electric and Hybrid Propulsion Division, within the Office of Transportation Technologies – directs the National Fuel Cells in Transportation Program. Other hydrogen programs are scattered throughout several federal agencies, including the Departments of Commerce and Defense.

The National Hydrogen Program

Since 1990, the national hydrogen program has been guided by the Sparks M. Matsunaga Hydrogen Research, Development and Demonstration Act.[2] Sponsored by the late Senator Matsunaga from Hawaii, one of the few longtime congressional hydrogen advocates, this law provided the first major federal impetus to consolidate, coordinate, and expand upon the small number of hydrogen projects scattered among federal agencies at the time.

The law designated the Department of Energy as the lead federal agency for hydrogen programs and charged it with three objectives: 1) to develop a five-year hydrogen research and development plan to "identify and resolve critical technical issues necessary for the realization of a domestic capability to produce, distribute, and use hydrogen economically within the shortest time practicable"; 2) "to develop a technology assessment and information transfer program among federal agencies" regarding hydrogen; and 3) "to develop renewable energy resources as a primary source of energy for the production of hydrogen." The Matsunaga Act further directed the Department of Energy to initiate research and demonstration projects for "critical technical issues... including use of hydrogen in surface transportation."

To advise the department in developing and implementing the national hydrogen research and development plan and to provide continuing guidance on hydrogen-related issues, the Act established a 12-member Hydrogen Technical Advisory Panel. The panel, which has met once or twice annually since its first meeting in February 1992, is divided into three subcommittees, including a Surface Transportation Subcommittee.[3] In 1993 it developed a "consensus statement" that refers to hydrogen as "the potential preferred energy carrier for transportation where high energy density is required."[4]

The Matsunaga Act authorized a total of $20 million over a three-year period to carry out the provisions of the law: $3 million for expenditures in 1992, $7 million in 1993, and $10 million in 1994. Federal government appropriations, however, did not match the authorized levels (see Table 1 on page 245). Program expenditures of $2 million in 1991 actually dropped after passage of the Matsunaga Act to $1.4 million in 1992. The budget more than doubled to $3.8 million in 1993, and in 1994 the $9.7 million appropriation nearly matched the level set in the Matsunaga Act. The National Hydrogen Program budget for 1995 edged up to $10 million. Despite the budget increases in the 1990s, the National Hydrogen Program is a very small effort within the total scope of federal energy research and development activities.[5]

The fundamental planning framework and program guidelines for the National Hydrogen Program, although broadened under the Clinton Administration, was first outlined near the end of the presidency of George Bush. The result was the 80-page *Hydrogen Program Plan, FY 1993 – FY 1997*, published in June 1992.[6] A more detailed strategy to execute the program is outlined in the *Hydrogen Program Implementation Plan, FY 1994 – FY 1998,* published in October 1993.[7]

Goals and activities

The goal of the National Hydrogen Program, pursuant to the Matsunaga Act, is to attain a level of hydrogen use in the United States between 0.1 and 0.5 quadrillion Btu by 2000; between 2.0 and 4.0 quadrillion Btu by

2010; and up to 10.0 quadrillion Btu by 2030. Total energy use in the United States is now about 85 quadrillion Btu, almost none of which is supplied by hydrogen. To achieve these goals, the National Hydrogen Program plan sets out activities in three areas: core research and development, systems analysis, and system demonstrations.

From its inception in 1992 through 1994, about 80 percent of the funds in the National Hydrogen Program have been allocated to the core research and development program, including hydrogen production, storage, and utilization. Systems analysis has accounted for the remaining 20 percent. Starting in 1996, however, demonstration projects, the third element, will begin to account for part of the program budget. The program had 15 funded projects in 1993.[8] The budget increase for the program in 1994 expanded this number to 31 projects at 15 universities, national research laboratories, and private-sector companies.[9]

- **Core research and development** Hydrogen storage projects represent the largest program element under the core research and development area, accounting for more than one-third of all program expenditures under the National Hydrogen Program. This effort's objectives are to develop hydrogen storage systems that: are at least 75 percent energy efficient, meaning only 25 percent of the energy in the hydrogen is lost on the way to, in, and from storage; that do not add more than $2 to $3 per million Btu to the cost of hydrogen (25 cents to 35 cents per equivalent gallon); meet a weight criterion of 4,000 Btu per pound of storage material; and meet a volume criterion of 2 equivalent gallons of fuel per cubic foot. Under this component of the National Hydrogen Program, hydrogen storage research funded in 1994 is underway at Syracuse University, the National Renewable Energy Lab (NREL), the University of Hawaii, the Florida Solar Energy Center, the Lawrence Livermore and Sandia National Laboratories, H Power Corporation, Energy Conversion Devices, and Arthur D. Little, Inc.

  Fuel production technologies represent the second component of the core research and development program area, targeted for about one-quarter of expenditures under the National Hydrogen Program.

Research projects addressing various hydrogen production processes – with a focus on biological and chemical photo-conversion technologies – are funded in 1994 at the Universities of Miami and Hawaii, Oak Ridge and Livermore National Laboratories, and NREL. (Hydrogen production technologies are discussed in Chapter 6.)

Hydrogen utilization projects make up the rest of the core research and demonstration program. Those funded in 1994 include three projects that relate directly to hydrogen engine development. This work is being done at three national laboratories: Lawrence Livermore, Sandia, and Los Alamos.

- **Systems analysis** Projects funded within the hydrogen systems analysis program element include studies of fuel cycle energy efficiency, safety, transportation and fueling infrastructure, and industry involvement in hydrogen development. Work funded in 1994 is taking place at Princeton University, the National Hydrogen Association, the University of Miami, the Florida Solar Energy Center, the Lawrence Livermore and Oak Ridge National Laboratories, the National Renewable Energy Laboratory, and Energetics, Inc.
- **Hydrogen demonstrations** The hydrogen demonstration project element of the National Hydrogen Program is scheduled to begin in 1997 with a project to build and test a hydrogen vehicle. The Department of Energy is planning later demonstrations to showcase hydrogen technologies in industrial and utility applications.[10]

In an effort to facilitate and increase private sector involvement in the various activities in the National Hydrogen Program, DOE initiated an Industrial Outreach Program in late 1994. A series of meetings, starting with a "super meeting" in April 1995, are designed to bring together companies active in hydrogen-related research and development to discuss ways to increase their participation in the national program and to identify promising opportunities for commercialization of hydrogen technology.[11]

Table 1. Funding for Major Federal Hydrogen Programs

Fiscal Year	National Hydrogen Program	Fuel Cells in Transportation Program
1991	$2.0 million	$8.9 million
1992	$1.4 million	$9.5 million
1993	$3.8 million	$12.0 million
1994	$9.7 million	$19.5 million
1995	$10.0 million	$24.6 million

National Fuel Cells in Transportation Program

The National Hydrogen Program initiated under the Matsunaga Act has been supplemented by and, in terms of federal dollars spent, surpassed by the National Fuel Cells in Transportation Program. The Office of Transportation Technologies at the Washington Department of Energy headquarters manages the program. DOE houses the program within the Office of Propulsion Systems rather than the Office of Alternative Fuels, owing to the early development status of hydrogen vehicles compared with vehicles powered by other alternative transportation fuels.

Although the Department of Energy has funded small-scale fuel cell engine projects for many years, it was not until 1993 that DOE prepared a comprehensive plan to promote the development of fuel cells for transportation applications.[12] Factors that contributed to the development of this plan included: the Clinton Administration's identification of hydrogen fuel cell technology as important to the national interest, clauses in the National Energy Policy Act of 1992 that require the development of both fuel cells for transportation and renewable hydrogen energy systems, and concern raised largely by the strides being made in Canada to develop the world's first fuel-cell-powered bus that the United States was falling behind its international competitors.

In February 1993, President Clinton's National Critical Technology Panel identified hydrogen fuel cells as of one 22 new technologies the development of which it considered essential to national economic prosperity and security. Total DOE funding for hydrogen fuel cell research for transportation has risen steadily in the 1990s, from $8.9 million in 1991 to $19.5 million in 1994, with a budget increase to $24.6 million in 1995 proposed by the Clinton Administration.[13]

Another reason for the development of the National Fuel Cell in Transportation Program is that fuel cell programs partially fulfill two requirements of the National Energy Policy Act of 1992.[14] Section 2025(j) of this Act directs the Department of Energy to "develop and implement a comprehensive program of research, development, and demonstration of fuel cells and related equipment for transportation… as the primary power source for private and mass transit vehicles and other mobile applications."

Moreover, Section 2026 of the Act directs the Department of Energy to conduct a five-year study of "renewable hydrogen energy systems… to supplement ongoing activities of a similar nature." Specific directives include demonstrations of "at least one program to develop a fuel cell suitable to power an electric vehicle" and another effort "to develop a hydrogen storage system suitable for electric motor vehicles powered by fuel cells."

As part of the Energy Policy Act directives, the Department of Energy published *National Program Plan: Fuel Cells in Transportation* in February 1993. The plan was developed with input from two groups: the Ad Hoc Technical Panel, including 19 hydrogen and fuel cell specialists from industry, academia, and government, and the five-member agencies and organizations of the National Fuel Cell Coordinating Group, consisting of the Energy and the Defense Departments, the Electric Power Research Institute, the Gas Research Institute, and the National Aeronautics and Space Administration.

Four activity areas

The National Fuel Cells in Transportation Program includes four components: engine development for heavy-duty applications, including buses and locomotives, fuel cell engine development for passenger vehicles, basic research and development of fuel cells and energy storage systems, and supporting analyses of such issues as safety and environmental impacts of fuel cell use.[15]

- **Heavy-duty engine development** The most ambitious of the four components is the Urban Transit Bus Program, co-funded by the United States Department of Transportation, the California Air Quality Management District, and a group of industries contributing 22 percent of the funding. The objective of this effort is to develop and demonstrate a bus powered by a phosphoric acid fuel cell and fueled by methanol reformed to hydrogen on board the vehicle. Phase I of the project, designing the bus, was completed in 1991. In Phase II, scheduled for completion in 1995, three prototype 30-foot buses will be built and road-tested. Future project stages will include the design and testing of small fleets of 40-foot fuel cell buses.

  DOE unveiled the program's first fuel-cell-powered transit bus in Washington, DC, as part of the Earth Day activities in April 1994. This bus was the first fuel cell bus built in the United States and, because it includes a large pack of batteries which are recharged by the fuel cell, it also represents the first fully integrated fuel cell electric hybrid bus in the world. The bus includes a 50-kilowatt phosphoric acid fuel cell and a methanol reformer with a combined weight of 3,000 pounds, plus 2,000 pounds of batteries and a 1,000-pound electric motor. This replaced a diesel engine and transmission weighing about half as much, or about 3,000 pounds. The batteries are stored underneath the bus, while the fuel cell and reformer fit into the existing diesel engine compartment. The bus's driving range between refuelings is about 150 miles while carrying 40 passengers.[16]

  H Power Corporation is the prime contractor for the fuel cell bus project, which includes the subcontractors Booz, Allen & Hamilton, Inc., Transportation Manufacturing Corporation, Fuji Electric, and Soleq Corporation. After being shown at conferences around the country in 1994, the bus is undergoing field testing by Georgetown University in Washington, DC. The Chattanooga Transit Authority, in Tennessee, will put a second fuel cell bus into service and the South Coast Air Quality Management District in Los Angeles will test a third bus in 1995.[17] A larger, 40-foot bus, also powered by a phosphoric acid fuel cell, is being developed in the next phase of the bus project.[18]

- **Passenger vehicle development** The passenger vehicle component of the National Fuel Cell in Transportation Program centers around the development of a proton-exchange membrane fuel cell small and light enough to be used in light-duty vehicles. The prime contractor for this effort was initially the Allison Gas Turbine Division of General Motors. GM, which subsequently sold its Allison Division, has remained involved, working with the Los Alamos National Laboratory, a main subcontractor, in New Mexico. Other subcontractors include Ballard Power Systems and Dow Chemical Company. As of mid-1995, Los Alamos National Laboratory was testing a 10-kilowatt proton-exchange membrane fuel cell. Future project phases include: scaling up the fuel cell to a 25-kilowatt size; later development of a commercially viable 50-kilowatt engine; and testing this engine in a prototype vehicle toward the end of the decade.[19]

- **Basic research and supporting analysis** Other projects underway as part of the National Fuel Cell in Transportation Program include studies of steam-reforming technology capable of converting methanol into high-purity hydrogen and technical assessments of a variety of onboard energy storage in addition to methanol. Arthur D. Little, Inc. of Cambridge, Massachusetts is the prime contractor for both projects. Moreover, the program is funding several research projects aimed at improving fuel cell performance as well as a number of technical support analyses. Finally, DOE initiated a small effort in 1993 to investigate the use of mixtures of hydrogen blended with natural gas as a fuel in internal combustion automotive engines.

Other Federal Initiatives Involving Hydrogen

Compared with the National Hydrogen Program and the National Fuel Cell in Transportation Program, much larger federal financial commitments are part of three other programs that potentially or actually involve applications of hydrogen in transportation: the Electric Hybrid Vehicle Program, run by the Department of Energy; the Technology Reinvestment Program, directed by the Department of Defense; and the Partnership for a New Generation of Vehicle, coordinated by the Department of Commerce. Although these programs do not focus primarily on hydrogen vehicle technologies, hydrogen fuel cells or hydrogen-powered internal combustion engines may emerge as a viable technology through their research agendas. All of these programs were established or significantly expanded in 1993, contributing to an all-time high level of federal involvement in hydrogen vehicle development. However, it is unclear whether this enthusiasm will continue in the face of competition from technologies such as electric batteries and advanced conventional engines, which are also represented in these programs.

Electric/Hybrid Vehicle Program

The DOE's Office of Propulsion Systems manages the Electric/Hybrid Vehicle Program, which supports the development of vehicles powered by a combination of power sources – including internal combustion engines, fuel cells, batteries, and electric motors. The budget for this program was $19 million in 1994, and the Clinton Administration proposed increasing it to $38.4 million in 1995.[20]

In the fall of 1993, the department chose the first two projects for funding under the Electric/Hybrid Vehicle Program. In each case, the Department of Energy will pay approximately half of all project costs.[21]

On September 30, General Motors and the Midwest Research Institute began a $138 million, five-year, DOE-funded project. The Institute, which operates the National Renewable Energy Laboratory in Golden, Colorado under contract to the Department of Energy, will represent DOE in this effort. The project's objective is to develop a hybrid vehicle equipped with both an electric motor and a conventional internal combustion engine.

In December 1993, the Department of Energy announced a second hybrid-vehicle development project. This $122 million contract includes Ford Motor Company and the Midwest Research Institute as key contractors. In October 1995, Ford and DOE allocated an additional $11 million to this project, specifically to study use of internal combustion engines in hybrid vehicles.[22]

In 1995, the Department of Energy and the major automotive manufacturers established the joint National Fuel Cells Alliance to coordinate fuel cell development efforts.[23] Also in 1995, the Department of Energy opened the Transportation Fuel Cell Information Center at its Energy Technology Engineering Center in Canoga Park, California. The Center maintains a database of publications related to fuel cell vehicles.[24]

Technology Reinvestment Project

The Department of Defense also conducts research programs involving electric and hybrid vehicle development through its Advanced Research Projects Agency (ARPA).

These programs are designed in part to implement President Clinton's Defense Reinvestment and Conversion Initiative established in early 1993. This initiative seeks to establish partnerships among federal government agencies, particularly the Defense Department, the 11 national research laboratories, the private sector, and other public-sector organizations.

The objectives of the initiative were incorporated into law through the Defense Conversion, Reinvestment, and Transition Assistance Act of 1993. To implement this law, ARPA established the Technology Reinvestment Project (TRP) in mid-1993. TRP aims "to stimulate the transition to a growing, integrated, national industrial capability which provides the most advanced, affordable military systems and the most competitive commercial products."[25] ARPA is the managing agency for the program, while overall program administration is a joint effort of the Defense Technology Conversion Council, which is chaired by ARPA and includes the Departments of Commerce and Energy, the National Aeronautics and Space Administration (NASA), and the National Science Foundation.

The federal budget for these activities in 1993 was $25 million, to be matched at least equally by the sponsors of projects receiving federal funds. In 1994, the budget was increased to $46 million.[26] In the first round of grants in 1993, ARPA awarded $23 million in support of six all-electric and hybrid-electric vehicle development consortia.[27] Each grant includes a two-year demonstration program involving prototype electric or hybrid electric cars or buses at an Air Force base. None of the 1993 grants specify development of hydrogen fuel cell vehicles, although development of hydrogen engines is likely to be included in several projects.[28]

The second round of projects funded by ARPA as of mid-1994 included 50 grants from 2,800 applications. Only one involves hydrogen energy. A $1.2 million grant was awarded to support a hydrogen vehicle demonstration project led by the Xerox Corporation and Clean Air Now, a public interest group.[29] The project is discussed in more detail later in this chapter.

Partnership for a New Generation of Vehicle

President Clinton announced the Partnership for a New Generation of Vehicle (PNGV), originally called the Clean Car Initiative, in February 1993. One of the partnership's key goals is to link and expand the research efforts at federal agencies with research conducted by the three major domestic auto manufacturers. The specific objective of the PNGV is to develop by 1998 a prototype automobile capable of achieving a fuel economy rating of 80 miles per gallon and to achieve commercial production of this vehicle within a decade. To meet these objectives, the initiative will, according to the White House, fund "research that could lead to production prototypes" demonstrating "radical new concepts such as fuel cells and advanced energy storage systems such as ultracapacitors."[30]

The program took a step forward on September 29, 1993, when President Clinton announced in a White House ceremony the establishment of a Master Cooperative Research and Development Agreement (CRADA) between the federal government and the Big Three auto manufacturers – General Motors, Ford, and Chrysler. The overall direction of the Partnership for a New Generation of Vehicle was assigned to the Department of Commerce, with other participating federal agencies including the Energy, Defense, and Transportation Departments, NASA, the Environmental Protection Agency, and the National Science Foundation. The automakers are represented in the CRADA through the United States Council for Automotive Research (USCAR), which was formed in 1984 subsequent to the passage of the National Cooperative Research and Development Act. This law authorized USCAR to coordinate joint research projects by automakers to develop precompetitive technologies without violating antitrust laws.[31]

The Master Clean Car CRADA consolidates and organizes research programs underway as part of four other CRADAs established earlier. They are:

- The Auto/Oil Air Quality Improvement Research Program
- The Environmental Research Consortium
- The Low Emissions Technologies R&D Partnership
- The Advanced Battery Consortium.

In September 1993, the Clinton Administration committed about $1 billion in federal funds to support the PNGV, with equal matching funds to come from private industry. No federal budget increases are involved, however; funds will be obtained by redirecting existing research programs. In early December 1993, the Big Three automakers signed a joint research agreement with the Department of Energy as an early step in implementing the PNGV. In January 1994, the first seven research projects, involving seven national research laboratories and the three automakers, were approved by the Clinton Administration with a total budget of $20 million spread over three years.[32]

Hydrogen fuels cells are among a wide variety of automotive fuels and technologies being investigated as part of the PNGV for potential use in its 1998 prototype car. In June 1994 testimony before the Senate Energy Committee, the director of the PNGV stated that as a result of "major breakthroughs," fuel cells would be among the technologies to receive focused attention in the PNGV. The director added that $60 million in contracts for fuel cell research were part of the first year's PNGV expenditures, including $13.8 million to Ford, $15 million to a Chrysler subsidiary, Pentastar, and $34 million to General Motors.[33] By the end of the first year's activities, more than $250 million in federal funds, two-thirds from the Department of Energy and the remainder from the other agencies in the partnership, had been allocated to contracts under the PNGV.[34] The proposed 1996 budget exceeds $300 million.[35]

Congressional Support for Expanded Hydrogen Programs

To date, the two laws that have done the most to promote the development of hydrogen vehicles have been the 1990 Matsunaga Act and the Energy Policy Act of 1992. Since the death of Sparks Matsunaga in 1990, leadership of hydrogen advocacy in the US Senate has passed to Senator Tom Harkin, a Democrat from Iowa. On December 3, 1992, Senator Harkin released his comprehensive plan to expand the federal government's renewable hydrogen programs. Harkin's plan would expand the National Hydrogen Program started under the Matsunaga Act to $135 million by the year 2000. Similarly, Harkin proposes to increase the National Fuel Cells in Transportation Program from the current $50 million in 1995 to $200 million by the turn of the century.[36]

In January 1994, Senator Harkin reiterated his plea for increased funding for the National Hydrogen Program and the National Fuel Cells in Transportation Program in a letter to the Assistant Secretary for Energy Efficiency and Renewable Energy at DOE. His letter outlined a proposal entitled "Sustainable Energy Centers – A Near-Term Path to Renewable Energy." These regional centers would implement hydrogen demonstration projects, train technicians and engineers in various hydrogen technologies, and educate the public about the potential advantages of hydrogen. While not abandoning the annual request to increase the budget for existing hydrogen programs, a process termed "incrementalism" by Harkin, the Sustainable Energy Centers are proposed to stake out "challenging but achievable goals to utilize hydrogen to help meet the President's energy and environmental objectives."[37]

In the House of Representatives, one of the strongest supporters of hydrogen energy development has been Robert Walker, a Republican from Pennsylvania who chairs the House Science, Space, and Technology Committee. On September 30, 1993, the day before the 1994 federal budget took effect, Walker introduced an amendment on the floor of the House of Representatives that passed, doubling the 1994 budget for the National Hydrogen Program to $10 million.[38] In March 1993, Rep. Walker introduced the Hydrogen Future Act, which authorized $100 million for hydrogen programs annually for five years: $60 million per year for hydrogen production and $40 million per year for demonstrations of end-use applications, including transportation.[39] The bill never reached the House floor for a vote in 1993 – an indication of the lack of interest in making hydrogen development a political priority.

In 1994, Walker reintroduced his hydrogen legislation, renamed the "Hydrogen and Fusion Research Authorization Act of 1994." It passed the House of Representatives on a voice vote in August 1994, but was never introduced in the Senate.[40] Rep. Walker returned again in 1995 to introduce the "Hydrogen Future Act of 1995," which authorizes $100 million in expenditures for hydrogen research and development over a three-year period.[41] The bill was passed by the House Science Committee in February 1995.[42]

Trade and Public Interest Organizations

A number of private industry and public interest organizations have appeared before Congress to endorse increased federal hydrogen energy activity. Some of the key players are:

- **Sustainable Energy Budget Coalition** This broad coalition of more than 100 environmental, energy policy, consumer, public health, government, and citizen groups leads the way in hydrogen advocacy.[43] In an attempt to influence Congress' energy budget priorities, the Coalition published a *Sustainable Energy Budget* in late 1992 and again in March 1994. One objective of the coalition, as stated in these publications, is to establish a national policy to reduce oil consumption in transportation by 50 percent within 20 years. Among the many specific initiatives supported in the proposed energy budget is government sponsorship of a "Green Machine" challenge aimed at the development of a pollution-free vehicle propelled by battery-stored electricity or by a hydrogen-powered fuel cell.

  Recommendations for the 1995 federal budget included increasing expenditures for renewable energy programs, which include hydrogen energy, to nearly double the $970 million requested by the

Clinton Administration. Specifically, the budget calls for an increase in the National Hydrogen Program budget from $10 million to $18 million and a doubling of all electric and hybrid vehicle activities to $147 million.

- **National Hydrogen Association** Based in Washington, DC, NHA is the leading trade organization for the hydrogen industry. Founded in 1989, its membership includes more than 35 private companies, including the major hydrogen producers and distributors, and several hydrogen consumers, most notably the National Aeronautics and Space Administration. The NHA has held annual hydrogen energy conferences in Washington each of its five years of existence through 1994. The fifth annual meeting, held in March 1994, drew about 200 participants, including more than a dozen representatives from Europe and Asia.[44]
- **American Hydrogen Association** The largest hydrogen advocacy group outside of Washington, DC, is the AHA, headquartered in Tempe, Arizona (AHA). Through congressional testimony, the AHA also endorses expanded hydrogen program activity by the federal government, with a particular focus on renewable hydrogen-production pathways and visionary end uses, including hydrogen-powered vehicles and all-hydrogen homes. On December 18, 1992, the AHA officially registered the nation's first two fully licensed hydrogen vehicles with the Arizona Motor Vehicle Department. The vehicles are a 1991 Oldsmobile, which is also capable of burning natural gas, and a 1979 Dodge pickup truck. The Association has also co-sponsored competitive road rallies of pollution-free hydrogen and solar vehicles and has hosted hydrogen energy conferences. As of the end of 1994, the AHA had chapters in several California cities as well as its lead group in Tempe.[45]

Several other trade organizations have recently formed to promote development of hydrogen fuel cell technology, including:

- **Fuel Cells for Transportation** Established in 1992, this is the largest group specifically encouraging use of fuel cells in automobiles.

About 30 organizations, mostly from California, are members of this Washington, DC-based lobbying organization. The group is seeking $450 million in federal funds over the next decade to support fuel cell research and development. Separate, but affiliated with Fuel Cells for Transportation, Fuel Cells 2000 was formed in 1994 to disseminate information about fuel cell technologies.[46]

- **American Fuel Cell Association** Together with its educational affiliate, the Fuel Cell Institute, this Washington, DC-based organization promotes all applications of fuel cells, with a particular focus on stationary applications. [47]
- **North American Clean Air Alliance** Formed in March 1994 by government agencies and companies primarily in California and British Columbia, Canada, the North American Clean Air Alliance supports the introduction of zero-emission vehicles, including hydrogen fuel cell vehicles.[48]

State and Local Government Hydrogen Initiatives

During the past few years, several state and local governments have funded projects to develop or test prototype hydrogen-powered vehicle technology, with a particular focus on fuel-cell and hybrid vehicles. Three states and the city of Denver have been most active in research and demonstration programs that include hydrogen vehicles, as discussed below.

Figure 2. Major State and Local Hydrogen Research & Development Activities

South Coast Air Quality Management District (CA)

Activity: **DOE-sponsored phosphoric acid fuel cell bus**

Activity: **Proton-exchange membrane (PEM) fuel cell**
Participant: Ballard Power Systems

Activity: **Production of solar-generated hydrogen, operation of refueling station, and vehicle demonstrations**
Participants: University of California at Riverside
Electrolyser Corp. Ltd.
Ontario Energy Ministry

Activity: **Pickup trucks running on solar-generated hydrogen**
Participants: Xerox Corp.
Clean Air Now
SEA Corp.
Praxair, Inc.
Advanced Machining Dynamics

Institute of Transportation Studies, University of California at Davis

Activity: **Proposed demonstration of five PEM fuel cell buses between Davis and Sacramento**
Participants: Sacramento Municipal Utility District
Ballard Power Systems
Science Applications International Corp.

City of Palm Desert (CA)
Activity: **Renewable hydrogen production and use in PEM fuel-cell-powered golf carts**
Participant: Humboldt State University

City of Palm Springs (CA)
Activity: **Demonstration of eight-passenger PEM fuel cell shuttle vehicle for use at regional airport**
Participants: Energy Partners, Inc.
Western Golf Car
Telesis Cogeneration

Denver Department of Health and Hospitals (CO)
Activity: **Demonstration of pickup trucks converted to run on natural gas mixed with hydrogen (Hythane)**
Participants: Hydrogen Consultants, Inc.
Public Service Co., of Colorado
Air Products and Chemicals Corp.

(more...)

Figure 2. Major State and Local Hydrogen Research & Development Activities (continued)

New York State Energy Research and Development Authority

Activity:	**Carbon adsorbent hydrogen storage technology**
Participant:	Advanced Storage Systems for Hydrogen Lab at University of Syracuse
Activity:	**Proton-exchange membrane fuel cell for hybrid electric bus**
Participant:	Mechanical Technology, Inc.
Activity:	**Proton-exchange membrane fuel cell for high-speed and low-cost manufacture**
Participant:	American Hydride Corp.

Pennsylvania Energy Office

Activity:	**Converted car to run with proton-exchange membrane fuel cell engine**
Participant:	Energy Innovations, Inc.
Activity:	**Tests of Hythane-powered van**
Participants:	National Fuel Gas Distribution Co. Air Products and Chemicals Corp. Hydrogen Consultants, Inc. Bruderly Engineering

California: State, Local, and Private Efforts

Since 1968, when Congress recognized the extreme air pollution problem in California, particularly in the southern region, the national Clean Air Act has permitted California to set automotive emissions standards stricter than federal standards applying to the other 49 states. California has repeatedly tightened air pollution control requirements for conventional vehicles and, for more than a decade, has also been the site of several of the world's most ambitious efforts to encourage use of alternative transportation fuels. The major players have been: state agencies that have funded alternative vehicle demonstrations and adopted strict regulations governing future automotive performance; local agencies, which grapple continuously with some of the world's worst air pollution problems; and several privately managed partnerships.

Low-emission vehicle program

At the center of California's alternative transportation fuels program are the low-emission vehicle regulations adopted by the California Air Resources Board in November 1990.[49] The regulations establish a four-stage reduction in vehicular air pollution emissions, to be phased in over a period of 15 years. The standards are so tight that full implementation will necessitate nothing short of an abandonment of the gasoline-powered vehicle in California. Starting in 1997, during the third stage, an increasing number of "ultra-low-emission vehicles" will be permitted to emit no more than 0.04 grams/mile of hydrocarbon emissions. This is 16 percent of the federal exhaust emission standard established in the Clean Air Act Amendments. Furthermore, starting with 2 percent of sales in 1998, "zero-emission vehicles"– defined as vehicles that release no pollution from the tailpipe – must be on California roadways.

The requirement for zero-emission vehicles is probably the most significant government action worldwide encouraging the development of electric vehicle technology. Battery-equipped or hydrogen fuel-cell-powered electric vehicles are the only currently available technologies that qualify under this definition.

Since 1990, hydrogen advocates have been trying to convince the Air Resources Board to designate internal combustion-engine-powered hydrogen vehicles as zero-emission vehicles. The key arguments supporting this approach are that hydrogen vehicles emit less pollution than battery-equipped vehicles on a total fuel cycle basis (from extraction through production and end use) and that, in any case, emissions from hydrogen vehicles are so low as to be environmentally inconsequential. There are some indications that the Air Resources Board may change its position. At a 1993 conference, for example, the board's Executive Director indicated that it was studying the possibility of defining a new emission category, called the near-zero-emission vehicle, to include hydrogen vehicles. Moreover, in 1993, the board approved the use of fuel-burning heaters on board battery-equipped electric vehicles designated as zero-emission vehicles. Emissions from heaters, which typically burn natural gas or propane, are greater than the discharges from hydrogen vehicles. This action, therefore, sets a precedent for qualifying hydrogen vehicles as zero-emission vehicles. In May 1994, the board convened public hearings to discuss possible changes to the regulations. But after two days of extensive testimony, the Board voted to retain the zero-emission vehicle requirement as originally enacted.[50]

On the other hand, a bill has been introduced into the California legislature that would overturn the zero-emission vehicle requirement. The bill, AB 2495, received support from the auto and oil industries at legislative hearings in March 1994.[51]

Meanwhile, hydrogen vehicles qualify as ultra-low-emission vehicles under the 1990 regulations. Even this designation is sufficient to stimulate interest in the development of hydrogen vehicle technology. Starting in 1997, two percent of all vehicle sales in California must meet the ultra-low emission standards starting in 1997; the percentage rises to 15 percent by 2003.

In another action regarding hydrogen at the state government level, the California Energy Commission funded a hydrogen vehicle assessment study, which was published in 1991.[52]

South Coast Air Quality Management District: four demonstration projects

At the local level, Los Angeles represents the center of activity regarding hydrogen vehicle development. The South Coast Air Quality Management District – the world's largest local pollution control agency, with 980 employees and an annual budget of more than $100 million – is the lead government agency sponsoring these activities. Several hydrogen projects are being managed as part of the district's Technology Advancement Office, including four hydrogen-vehicle demonstration projects underway in the Los Angeles area that receive partial funding from the District:[53]

- **Phosphoric acid fuel cell bus** The district has awarded $1.7 million in support of the Department of Energy's phosphoric acid fuel cell bus program, described earlier in this chapter. One of the three buses in this program will be field-tested in Los Angeles.

- **Proton-exchange membrane fuel cell bus** The district is a member of the British Columbia Fuel Cell Bus Steering Committee, which is developing proton-exchange membrane technology for buses under the direction of Ballard Power Systems (as described in Chapter 9). In late 1994, under the district's sponsorship, the first Ballard fuel cell bus was put into shuttle service at Los Angeles International Airport.[54]

- **Solar hydrogen research facility** The third project involves support for a multiyear program underway in Riverside, California, located in the eastern and most polluted portion of the Los Angeles basin. The project has proceeded on and off since the mid-1970s. In 1976, the internal combustion engine in a 19-passenger bus was converted to run on hydrogen and tested at Riverside Community College; in 1979, a second 25-passenger bus and a pickup truck were similarly converted to hydrogen.

  Hydrogen research at Riverside lapsed for more than a decade but was revived in the late 1980s when the district joined a private firm and a public agency from Canada in a project to produce hydrogen from solar energy. This effort includes the installation of a 3.5-kilowatt hydrogen electrolyser, which uses electricity generated by photovoltaic cells to split water and create hydrogen. The solar hydrogen system cost $600,000 in total, 75 percent of which was paid by the system's manufacturer – The Electrolyser Corporation of Ontario, Canada – and the Ontario Energy Ministry. The District contributed the remaining $150,000. In the second phase of the project, the District awarded a $180,000 contract to the Riverside Community College to convert two vehicles to run on the hydrogen produced by the solar system.[55]

  In 1992, the entire project moved to the nearby University of California at Riverside campus, where it has been incorporated as part of the College of Engineering's Center for Environmental Research and Technology (CE-CERT). The component of the project that includes the solar hydrogen production equipment is called the Solar Hydrogen Research Facility. In 1994, the facility added a fast-fill hydrogen vehicle refueling pump to the solar photovoltaic array and the electrolyser apparatus inherited from Riverside Community College. The entire system is the first totally solar-powered

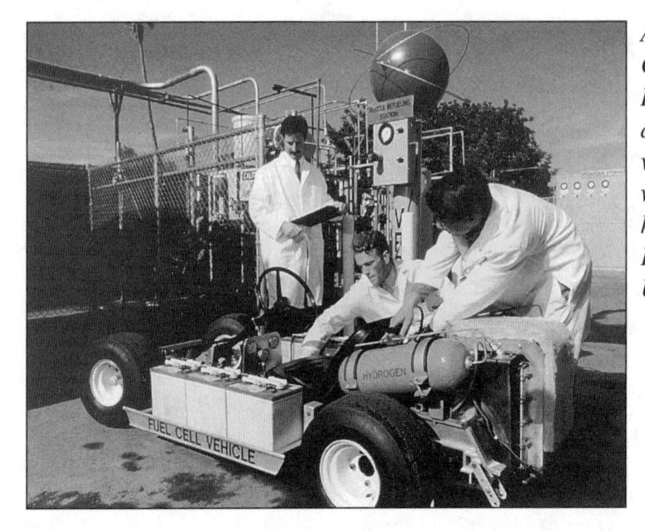

At the University of California at Riverside, demonstration vehicles are refueled with solar-generated hydrogen. Photo: CE-CERT, UC Riverside.

hydrogen vehicle refueling station in the country, and one of only two in the world (the other, in Germany, is discussed in Chapter 9).

CE-CERT's $10 million Vehicle Emissions Research Laboratory, located across campus from the Solar Hydrogen Research Facility, builds and tests hydrogen vehicles. As of March 1995, two hydrogen vehicles were undergoing technical evaluation. The first was a 1994 Ford Ranger equipped with an internal combustion engine converted to burn hydrogen. The second was a small, single-passenger racing car, not much bigger than a go-cart, equipped with proton-exchange membrane hydrogen fuel cells.[56]

- **Clean Air Now Solar Hydrogen Project at Xerox** The district has supported a project jointly sponsored by Xerox Corporation and an environmental organization, Clean Air Now – located across the Los Angeles Basin from Riverside in El Segundo. The project, initiated in mid-1993, seeks to build a solar hydrogen production facility at Xerox's El Segundo headquarters and to use the hydrogen to operate several Ford Ranger pickup trucks converted to run on hydrogen. By early 1994, the project's sponsors had raised $1.5 million for the project, including a $250,000 grant from the district and $1.2 million through the ARPA Technology Reinvestment

Project, described earlier in this chapter. The total project cost is estimated to be $2.5 million.[57]

A 50-kilowatt solar photovoltaic array, about 15 times larger than the array at Riverside, will generate the electricity to power the electrolyzers at the Xerox facility. The hydrogen produced by the electrolyzers will be used to refuel three hydrogen vehicles – 1993 Ford Ranger pickup trucks equipped with internal combustion engines – at a slow-fill refueling station. The entire solar hydrogen production and vehicle refueling project is scheduled for completion in 1995. In addition to Xerox and Clean Air Now, other project participants include SEA Corp., a builder of photovoltaic arrays; the hydrogen producer Praxair, which is constructing the refueling station; and Advanced Machining Dynamics, which will perform the vehicle conversions. United Technologies Corp. was originally designated to provide the electrolyzer, but the company withdrew in late 1994. Project sponsors are seeking additional corporate participants, including an electrolyzer manufacturer.[58]

Davis/Sacramento fuel cell bus fleet

Sacramento is another California city that has sponsored hydrogen vehicle research in part to address its environmental problems. In 1993, the Sacramento Municipal Utility District (SMUD) unveiled a $7.3 million, three-year Advanced & Renewable Technologies Development Program. As initially designed, the program included building one hydrogen fuel-cell-powered bus supplied with fuel produced at a 200-kilowatt solar-hydrogen production facility.[59]

The nearby Institute of Transportation Studies at the University of California campus in Davis is now leading the proposed bus component of the project. In early 1994, the Institute joined with Ballard Power Systems and Science Applications International Corporation to develop a Davis/Sacramento Hydrogen Fuel Cell Bus Demonstration Fleet. The proposed fleet is to include five buses, delivered by 1998, powered by proton-exchange membrane fuel cells. The buses would operate between Davis and Sacramento, with refueling facilities at the SMUD solar hydrogen facility in Sacramento and at a second facility to be built in Davis, 15 miles away.[60]

The Genesis Zero Emission Transporter runs on a proton-exchange membrane fuel cell. Photo: Energy Partners.

Palm Desert/Palm Springs hydrogen vehicle projects

Two small cities located in the desert expanses of southeast California – Palm Desert and Palm Springs – have initiated hydrogen vehicle demonstration projects. The Palm Desert project, begun in late 1994 and expected to cost $4.1 million, includes plans to use wind power to produce hydrogen by electrolysis. Palm Desert is to use the hydrogen to power golf carts equipped with 3-kilowatt proton-exchange membrane fuel cells (it is one of the few municipalities in the country to license golf carts for use on public roads). The Schatz Energy Research Center at Humboldt State University in Northern California is currently testing its prototype fuel cell golf cart.[61]

In nearby Palm Springs, Energy Partners, Inc., Western Golf Car, and Telesis Cogeneration have joined the city to build an eight-passenger, open-air shuttle called the Genesis Zero Emission Transporter. Powered by a 7.5-kilowatt proton-exchange membrane fuel cell, the Transporter can operate for three hours at 15 miles per hour between refuelings with compressed hydrogen. A prototype, completed in February 1995, is undergoing tests. The Palm Springs Regional Airport has committed to purchase and test five additional Transporters once they are built.[62]

Lawrence Livermore Lab: Electric hybrid project

In another California project that has expanded its work with funding from the National Hydrogen Program, Lawrence Livermore National Laboratory, in Livermore, about 30 miles east of San Francisco, is developing a design of a hybrid electric vehicle powered by a hydrogen-burning internal combustion engine coupled with an onboard electrical generator. The electricity produced can be used immediately to propel the vehicle or it can be stored in a battery, flywheel, or ultracapacitor system until needed. Initial studies suggest that this hybrid design can achieve energy efficiencies about equal to hydrogen fuel-cell-powered vehicles at a much lower cost.[63]

Denver, Colorado:
Hythane Demonstration

One of the oldest, continuous hydrogen vehicle demonstration programs is under way in Denver, Colorado. It is among the few projects in the world investigating the use of a fuel mixture of 95 percent natural gas and 5 percent hydrogen by weight (85 percent natural gas and 15 percent hydrogen by energy content), called Hythane. A consortium of public and private entities is testing the fuel in natural gas vehicles equipped with internal combustion engines modified to burn gaseous fuels.[64]

There are two major objectives to the Hythane project:

- To reduce air pollution to levels below those produced by burning natural gas, which already results in much less air pollution than gasoline-burning.
- To demonstrate a strategy for the logical transition from nonrenewable to sustainable energy resources, using natural gas as the intermediate step between gasoline and hydrogen.

The Denver Department of Health and Hospitals is managing the Hythane project, initiated in 1990, with additional funding from the US Department of Energy, state and local government agencies, and several private-sector participants. Hydrogen Consultants, Inc. developed and installed the technology to convert vehicles to burn Hythane. Other private-

Hythane allows this pickup truck to meet California's Ultra-Low-Emission Vehicle standard. Photo: Hydrogen Consultants, Inc.

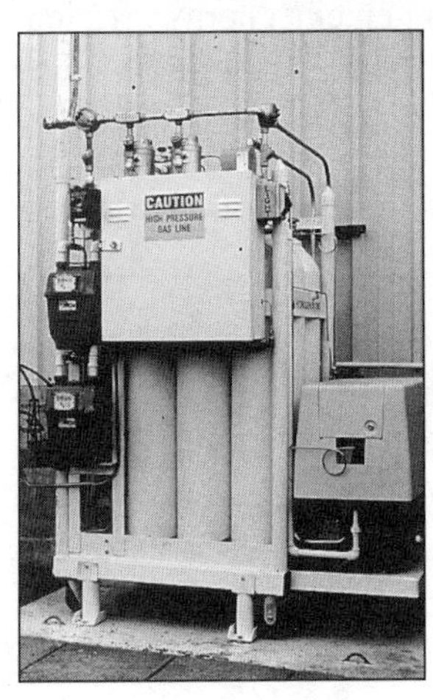

Fuel Maker's compressor blends natural gas and hydrogen to make Hythane for the Denver projects. Photo: Hydrogen Consultants, Inc.

sector participants include Public Service Company of Colorado, which contributes the natural gas and the natural gas vehicles, and Air Products and Chemicals Corporation, which contributes the hydrogen and the hydrogen storage equipment.

The first phase of Denver's Hythane project, completed in 1991, involved the conversion and testing of one Chevrolet S-10 pickup truck converted from natural gas to run on Hythane. Tests indicate satisfactory vehicle performance and a significant reduction in air pollution emissions over vehicles running solely on natural gas. These favorable results were confirmed in Phase II of the project, which included the comparison analysis of the performance of three vehicles: a conventional gasoline-powered S-10 pickup truck, a pickup converted to run on natural gas, and a pickup converted to run on Hythane. Phase III, begun in 1993, continues the comparative evaluation using three dedicated natural gas vehicles converted to run on Hythane.

Test results in 1993 fulfilled the project's first objective by indicating that the addition of 5 percent hydrogen to the natural gas can lead to 50 percent reductions in carbon monoxide and nitrogen oxide emissions, compared with emissions from burning pure natural gas. The tests also showed reductions of 30 percent in total hydrocarbon emissions. Although Hythane costs about 25 percent more than pure natural gas, the air quality benefits from using Hythane in polluted areas may justify the added cost. Project sponsors hope that the pollution reductions will provide the "leverage" needed to win a place in the alternative transportation fuels market for hydrogen, despite the high price of the fuel.

In keeping with the second objective of the Hythane project – to facilitate the transition from nonrenewable to sustainable energy resources – project sponsors also hope increased demand for hydrogen in Hythane fuel blends could stimulate interest in the development of hydrogen production technologies based on renewable energy resources. Such demand also could stimulate a gradual replacement of gasoline with natural gas and the subsequent introduction of hydrogen. To date, the Denver program has inspired the initiation of two other research and development programs to test Hythane, in Pennsylvania and Florida, discussed below.

New York: Alternative Fuels Research

The New York State Energy Research and Development Authority (NYSERDA) has been conducting an Alternative Fuels for Vehicles Demonstration Program since August 1990. The program has included funding of hydrogen storage and fuel cell research, but it is not demonstrating any hydrogen vehicles. The $40 million, five-year program is divided into two components. Vehicle demonstrations are by far the largest of the two elements. By the end of 1994, the Authority had funded an alternative transportation fuel fleet of more than 250 vehicles. About two-thirds of these vehicles are powered by natural gas and most of the remainder are powered by methanol. [65]

The second program component, research to accelerate the development of alternative fuel vehicle components, includes several hydrogen-related activities.[66] For several years, the authority has funded development of low-temperature carbon adsorbent hydrogen storage technology at the Advanced Storage Systems for Hydrogen Laboratory at the University of Syracuse. Funds from the authority are also supporting a number of hydrogen fuel cell research projects. First, Mechanical Technology Inc., in Latham, New York, is developing a proton-exchange membrane fuel cell. The objective is to produce a fuel cell for use in a hybrid-electric bus. The second project, under way at the laboratory of American Hydride Corporation in Hoosick Falls, New York, seeks to develop a PEM fuel cell that can be manufactured at high speed and low cost.

In early 1994, the New York State Energy Research and Development Authority began a program to fund electric-hybrid and electric vehicle projects, including hydrogen-powered fuel cell vehicles. NYSERDA awarded $500,000 in grants to support several other electric vehicle projects involving fuel cells.[67]

Pennsylvania: Fuel Cell and Hythane Projects

In 1988, the Pennsylvania Energy Office established its Alternative Transportation Fuel Program, which to this day remains one of the largest state government initiatives in the field. By the end of 1992, the program had awarded more than $2 million in grants to more than 30 projects. In late 1992, the legislature established a 0.5 percent tax on electric and natural gas utilities and earmarked the funds generated for use in funding alternative transportation fuel programs. This tax ensures that the Pennsylvania Energy Office will continue its involvement in encouraging development of alternative transportation fuel technologies.[68]

Like the New York program, most of the funds invested in Pennsylvania have supported demonstrations of natural gas vehicles, but the program has also supported two projects involving hydrogen vehicles.

- **Proton-exchange membrane fuel cell car** In 1991, The Pennsylvania Energy Office sponsored a demonstration of one of the earli-

Hydrogen Consultants' blender is part of the National Fuel Gas Hythane refueling station in Erie, PA. Photo: Hydrogen Consultants, Inc.

est fuel-cell-powered vehicles in the United States. In this project, Energy Innovations Inc. retrofitted a Ford Fiesta compact car with a proton-exchange membrane fuel cell. A $60,000 Energy Office grant funded Air Products and Chemicals to field-test this vehicle at its Allentown, Pennsylvania headquarters.

- **Hythane van** The second hydrogen project, initiated in 1993, involves the testing of a van powered by Hythane in Erie, Pennsylvania. The Energy Office contributed a grant of $59,700 to the project, which included the participation of the National Fuel Gas Distribution Company, Air Products and Chemicals Corp., Hydrogen Consultants, Inc. (the chief private-sector participant in the Denver Hythane project), and Bruderly Engineering. In 1993, the Erie Hythane project received the Pennsylvania Energy Office's Alternative Fuels Pioneer Award from the Energy Office.[69]

Private-Sector and Academic Programs

Perhaps the most startling evidence of increased interest in hydrogen vehicle technology in the United States has been the recent growth in private-sector activities, in large part spurred by the increase in federal and state funding. A survey completed by the National Hydrogen Association in March 1994 identified approximately 30 hydrogen research or demonstration projects underway in the United States, nearly all of which were directed by the private sector or universities.[70] Although many companies are participating at some level in these hydrogen projects, only a few companies and universities, mentioned below, are focusing on applications of hydrogen in vehicles.

Private-Sector Companies

- **H Power Corporation – The first American fuel cell bus** Founded in 1987 specifically to develop and manufacture hydrogen-powered fuel cells, H Power, in Belleville, New Jersey, is currently one of the companies most active in hydrogen vehicle projects. H Power is the prime contractor for the Department of Energy phosphoric acid fuel cell bus project, taking the lead in designing and building the nation's first fuel-cell-powered bus, which made its debut in April 1994. The company has invested about $3.6 million of its own resources in this project. The work on phosphoric acid fuel cells will continue until all three buses that are part of the DOE project discussed earlier are operating.[71]

  On its own, H Power is currently developing another fuel cell bus engine based on proton-exchange membrane technology. Commercial production is scheduled to begin in 1996. Designed as a retrofit package for existing diesel engine-equipped buses, the H Power bus will carry diesel fuel that will be reformed to produce hydrogen on board. Thus, the bus can be refueled using the existing diesel fuel infrastructure. The company believes that the weight of the entire power package – including the fuel cell, reformer, batteries, and motor – can be reduced to 2,500 pounds, about equal to the weight of conventional diesel engines and transmissions. The fuel cell itself will weigh about 500 pounds, or about one fourth as much as the phosphoric acid fuel cell in the other bus. In January 1995, H Power opened a manufacturing facility in Sacramento, California to produce proton-exchange membrane fuel cells. The $3 million factory will produce fuel cells in various sizes ranging from 25 watts to 10 kilowatts, with the larger units geared for use in hydrogen vehicles.[72]

H Power is also the prime developer of sponge-iron hydrogen storage technology in the United States. The company expects that sponge iron can effectively compete with liquefied hydrogen as the technology to transport hydrogen in oceangoing ships. Liquefaction costs would be eliminated, iron-carrying ships are much easier and cheaper to build than refrigerated vessels, and the iron can be regenerated and reused. Sponge-iron technology is also suitable for use in automotive vehicles.

- **Energy Partners, Inc. – The world's first hydrogen fuel cell car** Since 1989, Energy Partners, in West Palm Beach, Florida, has invested heavily in fuel cell vehicle development.[73] The company has completely redesigned the conventional automobile from bumper to bumper to create the world's first automobile powered solely by hydrogen used in a fuel cell engine. The project, initiated in 1989, produced its first design of a prototype vehicle in 1991, a two-passenger sports car. This preliminary model was powered by a 170-cell proton-exchange membrane fuel cell developed and built by US FuelCells Manufacturing, Inc., a subsidiary of Energy Partners acquired in 1991. The production cost of the hand-built prototype was roughly $180,000.

  Energy Partners' car has undergone significant additional development since 1991. The total work effort has cost over $3.5 million, most of which has been paid by Energy Partners' founder and chairman, John H. Perry, Jr. The South Coast Air Quality Management District has contributed $400,000 and the Department of Energy has added $200,000.

  The first demonstrated version of the "Green Car,™" unveiled in October 1993, is equipped with three 15-kilowatt proton-exchange membrane fuel cells. The vehicle is built mostly of lightweight composite materials, but fully fueled, it weighs nearly 3,000 pounds; the fuel cell engine itself contributes about 600 pounds. About 400 cubic feet of hydrogen is stored at 3,000 pounds per square inch pressure in compressed gas cylinders. The fuel cell engine and storage cylinder are located in the back of the Green Car, where they consume all of what would otherwise be cargo space. The car accelerates from zero to 30 miles per hour in about 10 seconds, achieves a top speed of 60 miles per hour, and has a driving range of 60 miles between refuelings. The Green Car emits no air pollution. [74]

Energy Partners, Inc., which employs about 20 people, plans to build a 50,000-square-foot fuel cell research center and to continue to develop fuel cell technologies for automotive and for stationary source applications. In February 1994, the Florida State Energy Office awarded the company a grant to design a proton-exchange membrane fuel cell manufacturing plant and to establish fuel cell research laboratories at two Florida colleges.[75] Energy Partners is also involved in the Genesis Zero Emission Transporter project described earlier in this chapter.

- **Hydrogen Consultants, Inc. – Hythane projects** From its participation in the early hydrogen vehicle demonstration programs in Northern California in the 1970s, Hydrogen Consultants, Inc., in Littleton, Colorado, has worked in the hydrogen field for more than 20 years. The company has participated in many projects converting internal combustion-engine-powered vehicles to burn hydrogen.[76] These include vehicle conversions in Riverside, California and the Hythane-fueled vehicles in Colorado and Pennsylvania mentioned earlier in this chapter.[77]

Academic Institutions

Until the recent upsurge in government and private-sector interest, analysis of hydrogen vehicles as a potential answer to the nation's growing transportation problems was almost solely the domain of academic researchers. Colleges and universities continue to perform some of the critical policy studies and basic scientific research central to the advancement of hydrogen as a transportation fuel. Several of the largest academic centers in the

United States addressing hydrogen energy not already mentioned in this chapter include the following:

- **The Institute of Transportation Studies** Established in 1986 at the University of California Davis Campus, the institute has prepared many in-depth transportation policy analyses and overviews of hydrogen vehicle technology.[78] Among its publications are *Hydrogen Fuel-Cell Vehicles* (1992), *Solar-Hydrogen Fuel-Cell Vehicles* (1993), *Alternative Transportation Fuels* (1989), and *A Comparative Analysis of Future Transportation Fuels* (1987). In 1993, the institute opened the Zero Emission Research Center specifically to study vehicle technologies, including hydrogen fuel cells, that emit little or no air pollution. Approximately 10 people work on energy policy studies or on development of zero-emission vehicle technology at the institute.[79]

- **The Florida Solar Energy Center** Established in 1974 by the Florida legislature as part of the state's University of Central Florida in Cape Canaveral, the center opened an alternative fuels research laboratory focusing on hydrogen energy technologies in 1992. The laboratory was built primarily with funds provided by the state of Florida and the US Environmental Protection Agency. The Department of Energy has also funded metal-hydride hydrogen storage technology research at the Center.[80]

  In 1992, the Florida Solar Energy Center began a program to test different Hythane mixtures in vehicle engines. Preliminary results from this project released in a January 1994 report show that when the concentration of hydrogen in natural gas was raised to between 28 percent and 36 percent, exhaust emissions dropped well below the California ultra-low-emission vehicle standard, even if the vehicle's catalytic converter was disconnected.[81]

Other academic centers of hydrogen research include the following:

- **The Clean Energy Research Institute at the University of Miami** The institute has been analyzing hydrogen energy systems for more than a decade. In recent years, it has produced several studies of safety aspects of hydrogen in comparison with other fuels. One study analyzed risks posed by hydrogen leaks in homes, assuming there was a home heating system, and a second examined dangers posed by hydrogen leakage from distribution pipelines. The institute offices have also served as the headquarters for the International Association for Hydrogen Energy since 1974.

- **The Center for Energy and Environmental Studies at Princeton University** The center has produced a number of economic and policy studies analyzing life-cycle costs and impacts of hydrogen technology.
- **The Center for Electrical Systems and Hydrogen Research at Texas A&M University** This group specializes in the development of lightweight proton-exchange membrane technology for transportation applications.
- **Schatz Energy Research Center at Humboldt State University** Since 1992, the Schatz Energy Research Center has been building a solar-powered hydrogen production facility in the Northern California seacoast town of Arcata. The solar generator is connected to an electrolyzer which produces hydrogen for use in a proton-exchange membrane designed and built by the center. The oxygen that is also produced by the electrolyzer is used to help aerate an aquarium on campus. The center is also involved in the Palm Desert project, described earlier. [82]
- **The Joint Center for Fuel Cell Vehicles** Founded in 1994 at the Colorado School of Mines in Golden, Colorado, the Joint Center for Fuel Cell Vehicles is seeking to develop partnerships of companies and organizations to perform specific hydrogen research projects. As of May 1995, two projects for which the center was seeking funding included the construction of an ammonia-powered fuel cell vehicle and the completion of a novel metal-hydride hydrogen storage system.[83]
- **National Laboratories** Most of the 11 government-funded National Laboratories throughout the United States perform some hydrogen

research. The four labs directly associated with the National Hydrogen Program in 1994 were Lawrence Livermore, Oak Ridge, Sandia, and Los Alamos National Laboratories.

- **Center for Hydrogen Energy at DOE's Savannah River facility** In 1994, the Department of Energy announced the formation of this hydrogen center at its longtime nuclear energy and weapons research facility in South Carolina. DOE proposes to make the Savannah River facility the manager of a project to build and test an urban transit bus powered by hydrogen burned in an internal combustion engine.[84]

Notes

1. A. Weisman, "Harnessing the Big H," *Los Angeles Times Magazine,* March 19, 1995.
2. Sparks M. Matsunaga Hydrogen Research, Development and Demonstration Act (P.L. 101-566), November 15, 1990.
3. "HTAP Subcommittee Call for Hydrogen Consortia," *The Hydrogen Letter*, October 1993; and "Pat Takahashi Elected New HTAP Chairman," *The Hydrogen Letter*, April 1994.
4. "Hydrogen Energy; Benefits, Applications and RD&D Needs: A Consensus Statement of the Hydrogen Technical Advisory Panel." (Washington, DC, June 7, 1993).
5. Senator Tom Harkin, "Sustainable Energy Centers: A Near-Term Path to Renewable Hydrogen" (Washington, DC, March 10, 1994, revised September 27, 1994).
6. United States Department of Energy, *Hydrogen Program Plan: FY1993 – FY1997* (Golden, CO, June 1992).
7. United States Department of Energy, *Hydrogen Program Implementation Plan, FY 1994 – FY 1998* (Washington, DC, October 1993).
8. National Renewable Energy Laboratory, *Proceedings of the 1993 DOE/NREL Hydrogen Program Review* (Golden, CO, May 4-6, 1993).
9. United States Department of Energy, *Participant Notebook for the Annual Review, Hydrogen Program Annual Review Meeting* (Livermore, CA, April 19-20, 1994).

10. N. Rossmeissl, "National Hydrogen Energy Program of the United States Department of Energy," *Hydrogen Energy Progress X: Proceedings of the 10th World Hydrogen Energy Conference* (Cocoa Beach, FL, June 20-24, 1994).

11. "DOE Starts Industrial Outreach Program," *The Hydrogen Letter,* January 1995.

12. United States Department of Energy, *National Program Plan: Fuel Cells in Transportation* (Washington, DC, February 1993).

13. Senator Tom Harkin, "Sustainable Energy Centers…," *op. cit.*

14. National Energy Policy Act (P.L. 102-486), October 24, 1992.

15. Thomas Gross and Pandit Patil, "Fuel Cell Development for Light-Duty and Heavy-Duty Vehicle Applications," paper presented by Steven Chalk at the *26th ISATA Conference* (Aachen, Germany, September 16, 1993).

16. J. Fisher, "Description and Test Results of the 30-foot PAFC Transit Bus," *Proceedings of the SAE Fuel Cell for Transportation TOPTEC Conference* (Santa Fe, NM, March 28, 1995); interview with Joseph Maceda, H Power Corporation, May 13, 1994; and interview with Sam Romano, Georgetown University, June 21, 1993.

17. "Energy Secretary O'Leary Rides H Power Bus," *The Hydrogen Letter*, May 1994.

18. J. Larkins, "The DOT/Georgetown 40-foot Advanced PAFC Transit Bus Program," *Proceedings of the SAE Fuel Cell for Transportation TOPTEC Conference* (Santa Fe, NM, March 28, 1995).

19. S. Swarthirajan, "General Motors Automotive Fuel Cell Program," *Proceedings of the SAE Fuel Cell for Transportation TOPTEC Conference* (Santa Fe, NM, March 29, 1995).

20. "Administration Proposes Increased Funding for Electric Vehicles," *Electric Transportation News*, March 1994.

21. *DOE News*, September 30, 1993; "Energy Department Contracts with GM Team for $130 Million Hybrid Development Effort," *Clean Fuels Report*, November 1993; "DOE Awards Hybrid Vehicle Contract to Ford," *Clean Fuels Report*, February 1994; and "DOE Awards Second Hybrid Contract to Ford," *The Hydrogen Letter*, January, 1994.

22. *Ford Motor Company News*, October 4, 1994, reported in *Alternative Energy On Line Today*; and "Ford Begins Work on Direct Hydrogen-Fueled PEM Fuel Cell for Transportation," *Clean Fuels Report*, February 1995.

23. "Big Three Turn Focus to Light-Duty Passenger Car Under National Fuel Cell Alliance," *Electric Vehicle On Line Today*, May 1995.

24. Written communication from Ken Sprouse, Transportation Fuel Cell Information Center, to INFORM, May 1, 1995.

25. Advanced Research Projects Agency, *Program Information Package for Defense Technology Conversion, Reinvestment, and Transition Assistance* (Washington, DC, March 10, 1993).

26. Presentation by Ellen Young, Electric Transportation Coalition, *5th Annual Texas Alternative Fuels Fair* (Austin, TX, April 18, 1994); and "ARPA Awards Defense Dollars for EV Programs," *Clean Fuels Report*, September 1993.

27. *Clean Fuels Report, ibid.*

28. The six recipients of the ARPA grants are: CALSTART in Burbank, CA ($4 million in ARPA funds) to develop 77 electric and 5 hybrid vehicles; The Sacramento Municipal Utility District ($2.5 million) for infrastructure development and flywheel and battery research; Hawaii Electric Vehicle Demonstration Consortium ($5 million) to test 55 electric vehicles; Mid-America Electric Vehicle Consortium in Indiana ($4 million) for unspecified projects; Southern Coalition for Advanced Transportation in Atlanta, GA ($4 million) to develop and test batteries, flywheels, and lightweight materials; Northeast Alternative Vehicle Consortium in Boston ($5 million) for electric vehicle technology development and testing.

29. "Xerox/CAN Plan: Defense Conversion to Solar Energy is Ready Now!," press release, Clean Air Now (Venice, CA, March 15, 1994).

30. President William Clinton and Vice President Al Gore, Jr., *Technology for America's Economic Growth: A New Direction to Build Economic Strength*, (The White House, Washington, DC, February 22, 1993); "White House Hi-Tech Plans Include Hydrogen, FCs," *The Hydrogen Letter*, March 1993; and "Clinton's Proposed Clean Car Development Effort Still Being Defined," *Clean Fuels Report*, June 1993.

31. "A New Partnership for Cars of the Future," press release (The White House, Washington, DC, September 29, 1993); "USCAR Broadens Its Scope of Interest," *Clean Fuels Report*, September 1993; and "Clean Car Project Gets Underway," *Clean Fuels Report*, February 1994; and "Washington Joins Big 3 Auto Venture," *The New York Times*, September 29, 1993 and "Government Dream Car," *The New York Times*, September 30, 1993.

32. "Clean Car Initiative Makes Progress," *Clean Fuels Report*, June 1994.

33. "DOE Fuel Cell Contracts with Big Three Imminent," *The Hydrogen Letter*, July 1994; and "PNGV Gives $29 million for Hydrogen-Fueled Auto Fuel Cells," *Clean Fuels Report*, September 1994.

34. "One Year Progress Report on Historic Partnership for New Generation of Vehicles," press release (The White House, Washington, DC, October 19, 1994); and P. Patil, "Fuel Cells for Transportation in the Context of the Partnership for a New Generation of Vehicles," paper presented at the *27th ISATA Conference* (Aachen, Germany, November 2, 1994).

35. "Funding for Government/Automaker Vehicle Partnership Debated as 1996 Budget Proposed," *Automotive News,* February 6, 1995, reported in *Alternative Energy On Line Today.*

36. Senator Tom Harkin, "Proposal for a Sustainable Energy Future Based on Renewable Hydrogen," (Washington, DC, December 2, 1992, revised June 3, 1993).

37. Senator Tom Harkin, "Sustainable Energy Carriers..." *op. cit.;* and "Sen. Harkin Proposes Two-Part Hydrogen Program," *The Hydrogen Letter*, February 1994.

38. "Congress Ups Hydrogen Budget to $10 Million," *The Hydrogen Letter*, October 1993.

39. "Rep. Walker Introduces New Hydrogen Bill," *The Hydrogen Letter*, April 1993.

40. "Hydrogen/Fusion Bill Introduced in Congress," *The Hydrogen Letter*, August 1994; and "Congress: Still Only $10 Million for Now," *The Hydrogen Letter*, September 1994.

41. "Rep. Walker Introduces $100 Million Hydrogen Bill," *The Hydrogen Letter*, February 1995.

42. "House Science Committee Reports Hydrogen Bill," *The Hydrogen Letter*, March 1995.

43. Sustainable Energy Blueprint Coalition, *Sustainable Energy Budget* (Washington, DC, November 1992 and February 1994); and *Sustainable Energy Blueprint* (Washington, DC, November 1992).

44. Interview, Robert Mauro, National Hydrogen Association, April 26, 1994.

45. American Hydrogen Association, *Hydrogen Today*, various issues (American Hydrogen Association, Tempe, AZ).

46. Interview, Robert Rose, Coordinator, Fuel Cells for Transportation, April 27, 1994; and "Recommendation to Execute an Agreement with the Ad Hoc Coalition on Fuel Cells for Transportation," South Coast Air Quality Management District (Diamond Bar, CA, June 12, 1994).

47. American Fuel Cell Association, "Prospectus" (Washington, DC,); and The Fuel Cell Institute, "Prospectus" (Washington, DC,) sent to INFORM, April 1994.

48. "Clean Air Alliance to Meet in Victoria, BC," *The Hydrogen Letter*, August 1994.

49. "Legislation Directory," *Clean Fuels Report*, June 1995.

50. "California Air Resources Board Sticks to ZEV Rule," *The Hydrogen Letter*, June 1994.

51. "Automakers Await Future of 'Anti-EV' Bill in California," *Autoweek*, January 10, 1994; and the "Calstart Executive Testifies: Clean Air Laws Spurring EV Innovation," *Business Wire*, March 21, 1994, both reported in *Alternative Energy OnLine Today.*

52. D. Swan and D. Smith, *Hydrogen-Fueled Vehicles: Technology Assessment Report* (California Energy Commission, Sacramento, CA, June 1991).

53. South Coast Air Quality Management District, information packet, sent to INFORM, July 1, 1993; and interview with Alan Lloyd, Chief Scientist, SCAQMD, April 27, 1994.

54. "Ballard Unveils World's First Commercial-Use Fuel Cell Bus at LAX," *Business Wire*, December 12, 1994, reported in *Alternative Energy Online Today.*

55. "Advance in Solar Hydrogen: The Riverside Solar Hydrogen Production Project: Preliminary Report," (1991) and "Technical Status Report on Electrolyser's Photovoltaic Hydrogen Production and Storage Technology," (1993), Electrolyser Corp. Ltd.; interview, Diana Harris, Riverside Community College, May 8, 1992; and *CE-CERT News*, August 9, 1993.

56. Interview, James Heffel; and tour of the CE-CERT Solar Hydrogen Research Facility and the Vehicle Emissions Research Laboratory, January 23, 1995; and *CE-CERT News*, August 1993 and June 1994.

57. "The Xerox/CAN Plan: Solar Hydrogen Project Executive Summary," Clean Air Now, Los Angeles, CA, 1994.

58. *Ibid.*; and "Praxair Considers Safety for Hydrogen Refueling Facilities," *Clean Fuels Report*, June 1994; and "United Technologies Pulls Out of H-Fleet Project," *The Hydrogen Letter*, November 1994.

59. " Pollution Credits – Are They Real?," paper presented by Bud Beebe, Sacramento Municipal Utility District, at the *4th Annual Hydrogen Conference* (Washington, DC, March 25, 1993); and interview, Bud Beebe, Sacramento Municipal Utility District, May 19, 1994.

60. Interview, Marshall Miller, Institute for Transportation Studies, May 18, 1994.

61. P. Lehman, "Palm Desert Fuel Cell Vehicle Project," *Proceedings of the SAE Fuel Cells for Transportation TOPTEC Conference* (Santa Fe, NM, March 29, 1995).

62. R. Ross, "Commercialization of Advanced PEM Fuel Cell Power Systems," *Proceedings of the SAE Fuel Cells for Transportation TOPTEC Conference* (Santa Fe, NM, March 29, 1995).

63. Interviews, Ray Smith, Lawrence Livermore Laboratory, May 6, 1994 and May 19, 1994.

64. James S. Cannon, *Paving the Way to Natural Gas Vehicles* (INFORM, Inc., New York, 1993), pp. 72-75; Venkat Raman, "Hythane – A Status Report," presented at *4th Annual Hydrogen Conference* (Washington, DC, March 25, 1993); "Hythane Wins EPA Award," *Clean Fuels Report*, June 1992; "Progress Reported in Denver Hythane Program," *Clean Fuels Report*, September 1992; and "Hythane Emissions Can Be 50 Percent Lower than Natural Gas," *Clean Fuels Report*, February 1994.

64. Cannon, *op. cit., pp. 72-75.*

66. "Alternative Fuel Vehicle Research, Development, and Demonstration," NYSERDA, August 9, 1993; and letter from Lawrence Hudson, NYSERDA, to INFORM, April 29, 1992.

67. "Energy Authority Pushes Hybrid-Electric and Electric Vehicles," press release, NYSERDA, February 18, 1994.

68. Cannon, *op. cit., pp. 72-75*; and *Pennsylvania Energy News*, Winter 1993.

69. "Pennsylvania Opens First Hythane Fueling Station in Erie," *Hydrogen Digest*, July/August 1993, reported by *Alternative Energy On Line Today*, September 1993; and V. Raman *et al.*, "Hythane – An Ultraclean Transportation Fuel," *Hydrogen Energy Progress X: Proceedings of the 10th World Hydrogen Energy Conference* (Cocoa Beach, FL, June 20-24, 1994).

70. Interview, Robert Mauro, National Hydrogen Association, April 27, 1994.

71. H Power Corp., information package, sent to INFORM, June 1993; and interview, Joseph Maceda, H Power Corp., *op. cit.*; and "H Power Corp. Involved in Broad-Range Hydrogen Development," *Clean Fuels Report*, June 1995.

72. "H Power Sets Up Shop in California," *The Hydrogen Letter*, January 1995.

73. Energy Partners, Inc., information packet and press clippings, sent to INFORM, April 1994; and "Energy Partners Unveils Pollution-Free 'EP Green Car™,'" *Fuel Cell News*, Winter 1994.

74. "Hydrogen-based Fuel Cell Power System Drives Green Car Into Clean Future," *South Florida Business Journal*, March 5, 1993; and "Green Machine: Pollution-free Car Could Get Test Spin Soon," *Florida Sun-Sentinel*, May 5, 1993; and "Energy Partners Restructures, Rolls Out Green Car," *The Hydrogen Letter*, November 1993; and "Energy Partners Unveils Green Car," *Clean Fuels Report*, February 1994.

75. "Energy Partners Receives Grant for PEM Fuel Cell Car Project," *Clean Fuels Report*, April 1994.

76. Hydrogen Consultants, Inc., information package, sent to INFORM, October 1993; and interviews, Frank Lynch, Hydrogen Consultants, Inc., June 22, 1994 and May 8, 1995.

77. Interview, Frank Lynch, Hydrogen Consultants, Inc., May 8, 1995.

78. "Hydrogen Research at UC Davis," *Hydrogen Today*, Winter 1994; and interviews with the staff of the Institute for Transportation Studies, Davis, CA, May 19, 1994.

79. Interviews, Marshall Miller and Manohar Prabhu, Institute of Transportation Studies, Davis, CA, May 18 and 19, 1994.

80. Interview, Kirk Collier, Florida Solar Energy Center, May 5, 1992; and presentations made during tour of the Florida Solar Energy Center, June 20, 1994.

81. R. Hoekstra and K. Collier, "Demonstration of Hydrogen Mixed Gas Vehicles," *Hydrogen Energy Progress X: Proceedings of the 10th World Hydrogen Energy Conference* (Cocoa Beach, FL, June 20-24, 1994); and "Hydrogen Research Under Way at Florida Solar Energy Center," *Clean Fuels Report*, April 1994.

82. P. Lehman and C. Parra, "Hydrogen Fuel from the Sun," *Solar Today*, September/October 1994; and interview, Peter Lehman, Schatz Energy Research Center, January 23, 1995.

83. Interview, Arnold Miller, Director, Joint Center for Fuel Cell Vehicles, May 15, 1995.

84. "Savannah Corp., Alaska, Highlight NHA Meeting," *The Hydrogen Letter*, April 1994.

Chapter 9

An Overview of International Hydrogen Vehicle Programs

Government-sponsored hydrogen research and development programs are underway in more than 30 countries. As in the United States, concern about oil dependence and environmental protection have prompted many countries to look for alternative energy resources. Japan and Germany's preeminence in this field reflects also a goal to develop new transportation technologies and energy resources to sell in the global marketplace – a goal that exists in the United States but has not been a national priority.

World leadership in hydrogen research, except for applications in space travel, has never been centered in the United States. Japan is currently the world leader, with a 20-year, multibillion-dollar hydrogen program called WE-NET (World Energy Network). In the 1990s, Japan's WE-NET program eclipsed the German program as the world's most ambitious hydrogen development effort. Although cutbacks have been instituted in Germany, mostly due to the high cost of political reunification, it still ranks number two, with annual expenditures of roughly half those of Japan.

Canada has probably the fourth-largest hydrogen program in the world, with the United States ranking third, and Italy, Norway, Turkey, Russia, and Switzerland maintaining significant programs. Between 10 and 20 hydrogen-powered cars and buses are now being tested in other countries, and plans are being implemented to place many more on foreign roadways. According to a review completed in 1994 by the International Energy Agency, 11 of its 23 member countries are engaged in hydrogen research, totaling $55.8 million in national government expenditures in 1994, including the $10 million program in the United States.[1]

Table 1. Highlights of International Hydrogen Vehicle Development

Daimler-Benz (Germany)

Mercedes 230 E equipped with metal hydride hydrogen storage system and hydrogen-burning internal combustion engine (1989).

Hydrogen-powered buses and refueling facilities (2 projects scheduled for 1997).

Mercedes 180 van powered by proton-exchange membrane fuel cell and fueled by compressed hydrogen (1994).

BMW (Germany)

Luxury 735iL sedans capable of burning either gasoline or liquefied hydrogen (1990 and 1994).

Musashi Institute of Technology (Japan)

Liquefied hydrogen-powered internal combustion vehicles, including a modified Nissan sports coupe (1990) and a Hino Motors refrigerated delivery truck (1993).

Mazda Motor Corp. (Japan)

Hydrogen-burning rotary engine vehicles equipped with metal hydride storage, including the HR-X2 4-passenger car and Miata sports coupe.

Ballard Power Systems (Canada)

World's first bus powered solely by proton-exchange membrane fuel cells fueled with compressed hydrogen (1993). Operates as part of the British Columbia Transit fleet.

Germany as a World Leader in Hydrogen Vehicle Research

Throughout the 20th century Germany has been the world leader in the research and development of hydrogen energy technology. The country's present-day hydrogen energy program, although scaled down somewhat from levels in the late 1980s, nevertheless remains among the most ambitious in the world. The German program includes hydrogen vehicle production and testing, basic research into the technology required for handling and storing hydrogen, and the development of renewable energy technology capable of producing hydrogen.[2]

History of German Hydrogen Vehicle Research

The world's first major commercial use of hydrogen in transportation occurred in Germany before and during World War II, and the country has been at the forefront of hydrogen energy development ever since. Rudolf A. Erren, a German engineer, designed and built many of the world's first hydrogen-powered vehicles, beginning with his first patent for a dual-fuel engine capable of burning both hydrogen and diesel fuels in 1928. The Erren engine was eventually installed in more than 1,000 cars and trucks, including models built by virtually every major German automotive manufacturer. The German government equipped a hydrogen-powered rail car and a submarine powered by hydrogen with Erren's engine. This phase of the German hydrogen program came to an abrupt end in the waning days of the second world war (see Chapter 4).[3]

The OPEC oil embargo of 1973 accomplished what eluded hydrogen advocates in 1945; it rekindled interest in hydrogen vehicle technology.[4] As a country importing more than two-thirds of its energy resources and dependent on oil for 60 percent of its primary energy supply at the time of the embargo, Germany was extremely vulnerable to disruption in oil supply and energy price increases. Through a multifaceted national energy strategy, oil dependence has since been reduced to less than 40 percent of primary energy supply.

Vulnerability to oil sparked German interest in alternative energy sources, including hydrogen. The Federal Ministry for Research and Technology (Bundesministeriums fur Forschung und Technologie, or BMFT) sponsored an alternative energy study in 1974. Entitled "New Fuels for the Road: Alternative Fuels for Motor Vehicles," the study identified hydrogen and methanol as two of the most promising long-term alternatives to reliance on oil. The report established a high research and development priority for hydrogen, and the German hydrogen R&D effort has continued with varying levels of financial support since then.

A branch of the BMFT, the German Aerospace Research and Development Institute (Deutsche Forshungs und Versuchanstalt fur Luft und Raumfahrt, or DLR), conducts the overall management of the federal hydrogen program. Founded in 1969, the DLR employs more than 4,400 people, mostly scientists and engineers, at five national research centers. It has an annual budget of about $300 million. The Solar and Hydrogen Energy Technology Center is part of the DLR's research complex on the campus of the University of Stuttgart.

Transportation fuels received a large boost in the national alternative energy program in 1979, when the German government embarked on a 10-year "Alternative Energy Sources for Road Transport" project. The hydrogen energy components of this project involved $20 million of federal funds plus comparable matching contributions from industry sources. The DLR coordinated most of the government-sponsored hydrogen activities initiated under this program in the 1980s. One early DLR project included development of a liquefied-hydrogen automobile that was shipped to the United States in 1982 for testing at the Los Alamos National Laboratory.

The most significant hydrogen-vehicle demonstration project undertaken and completed in Germany in the 1980s involved the building and testing of 10 Daimler-Benz vehicles.[5] Daimler-Benz, the Stuttgart-based parent company for Mercedes automobiles, has been engaged in hydrogen-vehicle research and development since 1973.[6] Most of its early work focused on storing hydrogen in metal hydride systems, with the goal of developing safe and efficient hydrogen storage technology for on-board vehicle use. In 1974, the company produced the world's first hydrogen vehicle equipped with a hydride fuel storage system. With the exception of a demonstration project in Berlin, most early research efforts concentrated on basic hydrogen engine research.

From 1984 to 1988, five Daimler-Benz 280 TE passenger cars and five 310 delivery vans were test-driven more than 350,000 miles, mostly on the streets of Berlin. One passenger car was used by a member of the Berlin Senate and the other four operated as medical on-call service vehicles. Daimler-Benz distributed the vans among five organizations: the Berlin Gas Works, the Red Cross, the Association for Efficient Energy Use, the St. John's Ambulance Service, and the Jewish Community in Berlin. In addition to the government institute, the DLR, project participants included 14 private-sector contractors and subcontractors.

Berlin Hydrogen Demonstration Project

Engines:

- The five passenger cars were dual-fuel vehicles capable of burning mixtures of gasoline and hydrogen. While idling and at slow speeds, the internal combustion engine burned predominately hydrogen; at faster speeds the fuel mixture was largely gasoline.
- The gasoline tanks were removed from the five vans and the engines were modified so that they were "dedicated" to burning hydrogen.

Fuel Storage:

- All 10 hydrogen vehicles were equipped with hydride fuel storage tanks composed of metal particles containing titanium, manganese, and vanadium.
- The hydride storage units were 25 times as heavy and 7 times as large as a gasoline tank.
- The tanks provided a 70-mile driving range on hydrogen.

During the four-year demonstration period, the program encountered no major safety problems, and driver acceptance was high. However, the vehicles experienced about a 15 percent loss in power when operating on hydrogen. Moreover, the hydrogen storage capacity of the hydride tanks deteriorated due to impurities in the hydrogen. As a result, the tanks had to be removed and repacked with hydride before the test period was one-third completed.

The Berlin field test demonstrated the viability of hydrogen as an automotive fuel in only slightly modified conventional vehicles operated in normal driving conditions and patterns. It established one of the largest databases of economic and performance information regarding hydrogen technology for vehicles powered by internal combustion engines. The project also helped to identify areas, such as fuel storage, where additional technological progress is necessary.

Current Hydrogen Vehicle Development Programs

The German government reassessed its commitment to hydrogen energy in the late 1980s in light of the Berlin demonstration and other smaller efforts conducted during the previous decade. In April 1988, a technical committee established by the BMFT prepared a plan entitled "Summary Report on Solar Hydrogen Energy Economy," which outlined an extended and enlarged hydrogen energy effort. Implementation of this plan resulted in sharply increased national government expenditures for hydrogen research and development. Hydrogen program spending rose from about $3 million in 1986 to a peak of $11 million in 1991, followed by three years of steady decreases, and reaching $8 million in 1994.[7] In 1994, federal hydrogen program expenditures in the United States exceeded those in Germany for the first time in more than a decade.

The BMFT is the sole source of financial support for most hydrogen research projects the DLR undertakes. Moreover, it provides up to 50 percent of the costs of many joint projects that include private-sector participants, including the hydrogen work of the country's two largest producers of high-performance automobiles, Daimler-Benz and BMW, described below. Among the world's major automotive manufacturers, only Mazda in Japan conducts hydrogen vehicle research on a scale comparable to these two German manufacturers.

Daimler-Benz initiatives

By 1989, Daimler-Benz's accumulated experience with hydrogen technology culminated in its most advanced prototype vehicle, a Mercedes 230 E. The company showcased this car at automotive exhibitions and energy conferences around the world.

- **Hydride storage for internal combustion engines** The prototype Mercedes 230 E, like all Daimler-Benz hydrogen automobiles produced to date, is equipped with a metal hydride fuel storage system. The metals attract and hold hydrogen, binding it to the solid through chemical bonds. The system consists of four tanks, that in total add more than 700 pounds to the vehicle weight while providing only enough fuel storage for 80 miles of driving. A liquid similar to antifreeze (composed primarily of glycol) is circulated through the hydride to absorb and remove heat released during refueling and to provide the heat necessary to release hydrogen from the hydride material as the car is driven. When fully fueled, hydrogen only accounts for about 1.3 percent of the weight of the fuel storage system; the rest is the metal hydride and the container.

In order to burn hydrogen, Daimler-Benz modified the fuel intake system to mix jets of hydrogen gas with air prior to injection into the engine cylinder. To eliminate engine backfiring that often results from such "external mixing" technologies involving hydrogen, small quantities of water are injected directly into the cylinder (as described in Chapter 4).

- **Liquefied hydrogen storage** Since 1990, Daimler-Benz has begun a shift in research priorities away from development of conventional hydride-equipped passenger vehicles to bus designs using liquefied hydrogen stored in cryogenic vessels. Moreover, engine research is focusing on direct injection of hydrogen into the cylinder, an approach called "internal mixing," to avoid backfiring associated with external mixing technologies. Part of the reason for the shift to liquefied hydrogen storage stems from the Euro-Quebec Project, discussed later in this chapter, which could provide liquefied hydrogen imports to Europe. The interest in bus technology also stems in part from the Euro-Quebec Project, which has identified urban transportation systems as promising niche markets for hydrogen vehicles.[8]

Two projects have been proposed to demonstrate the hydrogen bus technology being developed by Daimler-Benz. In late 1994, the Bavarian State Ministry of Economics and Transportation awarded a $220,000 grant to a Daimler-Benz subsidiary to design a hydrogen refueling system for installation at the three-year-old Munich airport. If the project moves ahead, the first hydrogen buses would begin operating at the airport in 1997.[9] Early in 1995, the city of Karlsruhe, Germany, announced a second project to demonstrate four hydrogen-powered buses built by Daimler-Benz. The local tran-

The fuel cell in the Mercedes 180 van was developed by the Canadian-based Ballard Power Systems. Photo: Hydrogen Consultants, Inc.

sit authority, Stadtwerke Karlsruhe, would operate the buses by 1997. The city's main utility company, Badenwerk, is currently directing the project and seeking co-sponsors to share the project's estimated $6.7 million cost.[10]

- **Development of fuel cells** In 1993, Daimler-Benz expanded its hydrogen activities to include fuel-cell-powered vehicles by joining an ongoing effort by the Canadian Ballard Power Systems, Inc. to develop a fuel-cell-powered bus.[11] Under the agreement, Daimler-Benz will share its considerable hydrogen expertise with Ballard Power Systems and invest $25 million over a four-year period toward the development of the fuel cell and associated components. The Ballard bus project is explained in more detail later in this chapter.

  Daimler-Benz has acted quickly to incorporate its recent experience in fuel cell development into a prototype automobile. In April 1994, the company unveiled its first fuel-cell automobile. The Mercedes 180 van, called the NECAR (for new electric car), is powered by a 42-kilowatt proton-exchange membrane fuel cell developed by Ballard and fueled by compressed hydrogen stored on board the vehicle. Based on several thousand miles of driving on the company's test track near Stuttgart, Daimler-Benz claims the performance of the car equals or exceeds the performance of gasoline-powered cars. The energy efficiency is approximately 30 percent, about twice the efficiency of conventional automotive drive trains. Even so, the vehicle range is only about 60 miles between refuelings. Daimler-Benz has suggested that it could begin commercial production of fuel-cell-powered subcompact automobiles as early as 1997.[12]

BMW: Liquefied hydrogen storage

Munich-based BMW has matched the efforts of Daimler-Benz, one of its chief competitors, step for step.[13] Its own hydrogen vehicle program dates from 1979. The BMW program has concentrated on developing storage technology for liquefied hydrogen, in contrast to the historical emphasis on hydride technology by Daimler-Benz. BMW has built and tested six generations of hydrogen vehicles, all of which have been fueled by liquefied hydrogen.

The last two prototypes, unveiled in 1990 and 1994, are modified 735iL luxury model sedans that are capable of burning either gasoline or hydrogen. The 3.5-liter engine is supplied with fuel through an external mixing process. Backfiring is controlled by maintaining a surplus air supply throughout the mixing and fuel delivery process. The use of excess air, called lean-burn technology, increases engine efficiency as well as controlling backfiring; however, it inhibits acceleration and peak speed, thereby increasing power losses, which can exceed 30 percent in the BMW vehicle when it is burning hydrogen. [14]

The liquefied hydrogen storage tank on the BMW 735iL, located in the trunk, provides enough fuel for about 180 miles of driving. Refueling with hydrogen takes about one hour, nearly all of which is spent connecting the refueling hardware to the vehicle and cooling the entire mechanism to prevent flash evaporation of liquefied hydrogen as it passes through warmer refueling hoses. The vehicle tank and interior are equipped with hydrogen detectors capable of sensing very small leaks in the fuel storage and delivery system. If a fuel leak is detected, the sun roof and windows automatically open to increase ventilation.

In June 1994, BMW unveiled its most recent prototype hydrogen vehicle before 300 journalists at a media event at its Munich headquarters.

The company also announced its intention to promote its hydrogen vehicle technology worldwide by shipping the new car for demonstration around the world, including extended demonstrations starting in 1995 in the United States and Canada.[15]

Since March 1989, BMW has been developing and testing hydrogen engines and automotive components at the world's first indoor testbed for hydrogen-drive engines. This test facility, located in Munich, is used exclusively for research and development of hydrogen engines. A 750-gallon liquefied hydrogen storage tank supplies the facility with fuel.

Like Daimler-Benz, BMW is participating in a number of national and international hydrogen projects (discussed later in this chapter). Also, most of BMW's hydrogen activities, like those at Daimler-Benz, have included participation of the DLR. For example, much of the development work for liquefied hydrogen storage systems suitable for automotive applications was performed by the DLR before it was applied in the BMW vehicles.

The SWB Solar Hydrogen Demonstration Project

In addition to the hydrogen vehicle development work in Germany, a number of private- and public-sector organizations are conducting several large solar-hydrogen production projects. Taken together, these projects offer some of the world's most impressive demonstrations of state-of-the-art technology for generating hydrogen from water using a clean, renewable resource. Two of these efforts, the Hysolar project, a joint effort involving Germany and Saudi Arabia, and the multinational "Euro-Quebec Hydro-Hydrogen Pilot Project" are discussed later in this chapter in the section on multinational projects. The exclusively German Solar-Wasserstoff-Bayern project is examined here.

Europe's largest solar-hydrogen production facility uses photovoltaic cells to power an electrolysis unit that separates hydrogen from water. The facility opened in October 1990 at Neunburg vorm Wald, about 120 miles northeast of Munich.[16] It is owned and operated by the Solar-Wasserstoff-Bayern GmbH (SWB), a joint venture established in 1986. Partners in SWB include the German private utility company Bayernwerk (60 percent),

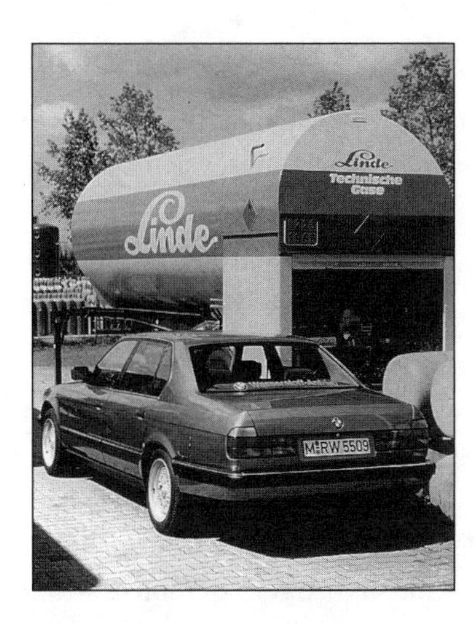

The SWB solar hydrogen project in Germany includes a liquefied hydrogen refueling station for the BMW demonstration sedan. Photo: BMW

BMW (10 percent), Linde, a producer of industrial gases (10 percent), Siemens, a major German electronics producer and a manufacturer of solar equipment (10 percent), and MBB, an aerospace company (10 percent). Project expenses are divided evenly between SWB and German government agencies. The federal BMFT provides roughly two-thirds of government funding, while the Bavarian state government supplies the remaining third.

The first phase of this large solar-hydrogen demonstration project, from 1986 through 1991, cost about $38 million. The SWB facility constructed during this phase includes 270 kilowatts from solar photovoltaic cells connected to hydrogen electrolysis units. Phase I also included a liquefied-hydrogen vehicle refueling station completed in April 1991. Since 1991, the BMW hydrogen demonstration automobile has travelled between the Munich BMW headquarters and the SWB facility. Refueling occurs at each stop. Initially, refueling at the SWB facility took about one hour. During the process, up to 30 percent of the liquefied hydrogen evaporated before it could be stored onboard the car. Since then, advancements in the refueling technology and improved procedures have reduced the time to 20 minutes and cut fuel evaporation to 10 percent.

Phase II of the project, scheduled to last through 1996 at an additional cost of $41 million, includes expanding the production capacity by adding more solar photovoltaic panels and constructing additional end-use test centers for stationary and vehicle use. In Phase II, the liquefied hydrogen refueling station will be expanded to service hydrogen buses and trucks as well as the BMW automobile.

Less Government Involvement Since 1991

During the late 1980s and early 1990s, the German government invested tens of millions of dollars in hydrogen research and development projects. This earned Germany its position as the world's most aggressive pursuer of a hydrogen energy future. The current hydrogen vehicle programs at Daimler-Benz and BMW, the SWB solar-hydrogen demonstration, Germany's involvement in several international hydrogen programs, and the continuing work of the federal DLR all combine to maintain German preeminence in hydrogen technology.

However, the high costs associated with the unification of Germany, which began in 1990, coupled with the low price and steady availability of oil, have recently undercut the German government's willingness and ability to maintain its high profile in hydrogen research and development. A September 1993 federal government cabinet report suggested that the high cost of hydrogen was likely to constrain government efforts to commercialize hydrogen technologies in the current economic climate.[17] Research and development activities continue, but with reduced funding. The federal hydrogen program budget in 1994 was $8 million, and the 1995 budget is $12 million – the second-highest in the world, after Japan.[18]

Hydrogen Programs in Japan

The Hydrogen Energy Systems Society was founded in Japan in 1973, initiating what has grown into one of the world's largest efforts to develop and apply hydrogen technologies.[19] Since 1973, the Japanese hydrogen program has been coordinated as a key component of "Project Sunshine," a national program to develop energy technologies based on renewable energy resources. The Ministry of International Trade and Industry (MITI) manages Project Sunshine, and its chief operating agency, the New Energy Industrial Technology Development Organization (NEDO), largely implements the project. NEDO, founded in 1980 after the major global oil price hike in 1979, has a staff of more than 800 people, 250 of whom work specifically on energy. Annual funding has ranged from 5 percent to about 20 percent of MITI's entire research and development budget and up to about 50 percent of the ministry's energy research and development budget. For most of the 1980s, the budget for Project Sunshine averaged about $300 million per year. Although less than $10 million of these funds were directed annually to hydrogen programs, Project Sunshine nevertheless represents one of the largest sustained efforts to develop hydrogen energy systems.

Moreover, in the 1990s, the national hydrogen effort has expanded significantly. Japan's expenditures of $19 million in 1994 and $23 million in 1995 are the largest of any country in the world, double those of its closest rival, Germany. In fact, according to an International Energy Agency survey of hydrogen programs in member countries of the Organization of Economic Cooperation and Development, in 1995 Japan accounted for close to half of the $55.8 million in hydrogen expenditures among 11 countries that support hydrogen research and development.[20]

Project Sunshine has funded research projects involving all aspects of hydrogen production and use, including:

- Development of hydrogen production technologies at the Government Industrial Research Institute of Osaka since 1974.
- Hydrogen storage research, primarily the development of metal hydride storage systems, at the National Chemical Laboratory for Industry.
- Hydrogen vehicle technology research at the Musashi Institute of Technology and by Mazda Motor Corporation.

Japanese Hydrogen Vehicle Projects

The Musashi Institute of Technology in Tokyo has been developing hydrogen vehicles since 1970. The institute has built and tested a total of

Two Musashi Hydrogen-Burning Vehicles

Musashi-8 sports car:

- Contains a 100-liter liquefied hydrogen tank, only 2.7 times the weight of a similarly sized gasoline tank. Musashi claims that this is only one-tenth the weight of a comparable metal hydride or compressed hydrogen storage cylinder.
- High-pressure injection of cold hydrogen, easily achievable with liquefied hydrogen systems, nearly eliminates the problem of backfiring.
- Powered by a 4-cylinder, 4-stroke, 100-horsepower diesel engine equipped with a spark ignition system.
- The two-passenger car, based on a 300 ZX sports coupe manufactured by Nissan Motor Company, is capable of a sustained speed exceeding 75 miles per hour.

Musashi-9 refrigerated delivery truck:

- First hydrogen-powered truck to be built in Japan, and among the first trucks in the world to be built and "dedicated" to burn only hydrogen.
- Equipped with a 400-liter stainless steel liquefied hydrogen storage tank that weighs about 900 pounds.
- The tank carries the equivalent of 30 gallons of diesel fuel and provides a 375-mile driving range for the truck.
- The 4-ton truck, based on a truck model built by Hino Motors Limited, a subsidiary of Toyota, is propelled by a modified 160-horsepower diesel engine.
- The liquefied hydrogen passes through a heat exchanger on the way to the engine. The cold air, generated in the exchanger at no extra cost, is used to refrigerate the cargo hold. The truck is thus perfectly suited to carry fruits, vegetables, meats, and other products which must be kept cool on the way to markets.

nine prototype hydrogen-powered vehicles, all powered by internal combustion engines. Two are currently on the road, the Musashi-8, a sports car shown first at the 1990 World Hydrogen Conference in Hawaii, and the Musashi-9, a refrigerated produce delivery truck unveiled at the New Energy Systems Conference in Yokohama in mid-1993.[21] Use of liquefied hydrogen has been a trademark of the Musashi hydrogen vehicles; the

Institute has pioneered the development of low-temperature hydrogen pumps and other fuel delivery components applicable in automobiles.

Mazda has also made a significant investment in the development of hydrogen automobiles in Japan.[22] Mazda unveiled its first hydrogen vehicle, dubbed the HRX, for "hydrogen rotary research," at the Tokyo auto show in 1991. The futuristic droplet-shaped car was built largely of lightweight recyclable plastics. In 1993, Mazda showed two new hydrogen prototype vehicles, the HR-X2 four-passenger car and a two-passenger Miata sports coupe. These cars have been widely shown at major auto exhibitions in Japan, Europe, and North America. The vehicles are part of Mazda's $35 million hydrogen research program that aims to begin commercial production of hydrogen vehicles as early as 1997.

Mazda has used metal hydrides to store hydrogen in its vehicles instead of the liquefied-hydrogen storage systems used in vehicles produced at the Musashi Institute. The hydride system in the Miata weighs 660 pounds, compared with a 110-pound conventional gasoline tank, yet it offers a driving range of only 100 miles, compared with more than 300 miles for the gasoline version. The hydrogen is burned in Mazda's unique rotary engines rather than in conventional piston-driven internal combustion engines used in the Musashi vehicles. The company asserts that the rotary engine eliminates backfiring when hydrogen is burned. Equipped with the same engine that powers the popular RX-7 sports car, the cars can reach a top speed exceeding 80 miles per hour.

The New Sunshine and WE-NET Programs

In the Spring of 1993, the Japanese government announced a significant rebirth of the Sunshine Project, called the "New Sunshine Project." This effort is budgeted at $11 billion over a 28-year period. The first-year appropriation was $390 million; the budget in 1994 was $497 million.[23]

A key component of the New Sunshine Project is the "WE-NET Project," budgeted for $2 billion of the $11 billion overall program. The overall objective of WE-NET ("World Energy Network"), is to develop and commercialize advanced energy technologies capable of reducing global carbon dioxide emissions by 10 percent in 2010 and by 50 percent in 2050. The "International Clean Energy Network Using Hydrogen Conversion" program constitutes a major component of the WE-NET effort. The project description states that the program aims at "constructing a global energy system which accords with ecosystem principles... while also creating new industries related to hydrogen."

- **Phase I, Research** To be completed during a four-year period from 1993 through 1996, this phase has a total budget of about $100 million and includes a variety of projects involving the production, storage, and use of hydrogen. Hydrogen production research will focus on electrolysis technologies; liquefaction and transportation of liquefied hydrogen by sea will dominate the agenda of hydrogen storage research; and hydrogen-use research will address applications in stationary and mobile engines, including hydrogen vehicles.
- **Phase II, Demonstration** The second phase will include the construction of facilities in Japan to show a variety of technologies to produce, store, and use hydrogen.
- **Phase III, Use** A fully integrated, international project for the production, distribution, and use of hydrogen will be completed as part of Phase III. In February 1995, NEDO staff working on the WE-NET program proposed a project, the "Trans-Asian Gas Pipeline Network," that could form a major component of Phase III. The 31,000-mile pipeline would initially be built at a cost of $92 billion to carry natural gas between 19 Asian natural-gas-producing regions and the major energy-consuming centers, forming an "Asian Pacific Energy Community." The pipeline would be made of high-tension steel, which has a lifetime of more than 100 years and is capable of transporting hydrogen safely as well as natural gas. As natural gas reserves deplete in the next century, hydrogen would be substituted in the pipeline.[24]

The advent of the New Sunshine Project, and specifically the WE-NET project, suggests that consistent, large-scale government funding of hydrogen energy projects in Japan will continue for years to come.

Hydrogen Programs in Canada

On a per-capita basis, Canada produces and consumes more hydrogen than any other nation on earth.[25] Most hydrogen activity historically has centered around Canada's large fossil fuel industries. Canada has extensive reserves of low-quality oil, coal, and tar sands. These energy resources can be upgraded to higher-quality fuels by adding hydrogen during fuel processing and refining. Hydrogen increases the energy content of the fuel and reduces the pollution when the upgraded fuel is burned. Hydrogen produced by steam reforming of Canada's abundant natural gas reserves has been used extensively for this purpose. Steam-reformed hydrogen on its own is more than a $1 billion per year industry in Canada. The country also contains enormous hydropower resources that offer a large alternative source of energy for hydrogen production in the future by electrolysis of water.

A major focus of the Canadian program has been to increase hydrogen production from its vast natural gas and hydropower resource base to serve world markets. Reduction in oil dependence, a main driving force elsewhere, has been a secondary concern. Development of technology to use hydrogen has been another key element of the hydrogen program. A strong national effort today is bolstered by major hydrogen projects underway in three Canadian provinces.

The Canadian National Hydrogen Program

The Canadian government and industry began to invest substantially in a national hydrogen program in the 1980s. In September 1985, the Ministry of Energy, Mines and Resources and the Ministry of State for Science and Technology jointly commissioned an Advisory Group on Hydrogen Opportunities. In June 1987, the nine-member Advisory Group issued its final report entitled *Hydrogen: National Mission for Canada.*[26] The report recommended that the federal government establish a national priority for Canada to become a world leader before the end of this century in the production of hydrogen and the development of technologies capable of using hydrogen.

In 1988 the federal government endorsed the Advisory Group's findings and directed the Hydrogen Industry Council to develop a national research and development plan for Canada. Founded in December 1982, the Hydrogen Industry Council was one of the first trade associations in the world formed specifically to promote development of hydrogen technologies and their use in industrial and commercial applications. Funded initially by a $1 million, five-year government grant, the Council has been sustained since 1988 by dues from its members and by government grants for specific projects.[27]

In March 1989, the Hydrogen Industry Council submitted a National Strategic Technology Development Plan that recommended a $150 million hydrogen research and development program, jointly funded by industry and government over a 10-year period. In March 1990, the Council followed up with a detailed two-year menu of projects.[28] This initial $2 million program, funded equally by the federal government and industry, was administered by the council. Major projects involved development of improved electrolyzers, advanced hydrogen-liquefaction technology, and high-efficiency fuel cells.[29] Federal government involvement in hydrogen programs has continued since the first two-year effort ended in 1993. Federal funds are included in most of the Canadian projects discussed in this section. Moreover, the federal government is participating in the International Energy Agency Hydrogen Agreement and the Euro-Quebec Project discussed later in this chapter.[30]

Provincial Hydrogen Projects

Several provincial governments in Canada have been active in hydrogen projects.

Ontario

The longest-standing hydrogen program is in Ontario; it began in 1980 with the work of the Ontario Hydrogen Task Force. The Ontario government has remained active by funding specific hydrogen projects undertaken by local industries and universities, including a major research project at ORTECH International, a large energy consulting firm, to design and build engines powered by Hythane. Other projects funded by Ontario include development of electrolyzer technology powered by solar energy.[31]

Quebec

The provincial government of Quebec is a partner in the Euro-Quebec Project discussed below. As part of the demonstration phase of this project, several buses will be tested on Montreal streets. The first Hythane buses, powered by a mixture of 20 percent hydrogen and 80 percent natural gas, were unveiled on November 25, 1994. The Canadian urban bus manufacturer Novabus built the 40-passenger, 40-foot urban buses, which the local transit authority operates in Montreal. The buses are powered by a Cummins L-10 engine, originally designed for diesel fuel but recently redesigned to burn natural gas. Most of the natural gas buses in the United States and Canada currently use the L-10 engine. The compressed gas Hythane mixture is stored on the roof of the bus in four lightweight tanks that hold enough fuel for a 250-mile driving range. A second Hythane bus took to the streets of Montreal in February 1995. The total cost of the Montreal Hythane bus project is about $4 million, more than 75 percent of which is contributed by the Quebec Ministry of Natural Resources. The remainder is being contributed by various industrial partners.[32]

Alberta

About 55 percent of Canada's hydrogen production and 63 percent of its consumption occurs in Alberta. Not surprisingly, Alberta has a large program aimed at improving hydrogen production and its use by its chief consumer, the Alberta petroleum industry. In 1988 the Alberta Energy Department established the Hydrogen Technology Research Program. Through 1992, the program had funded 20 projects, mostly involving technologies to produce hydrogen from fossil fuels and to use it to improve the quality of Alberta's coal and oil resources.[33] In 1993, the Energy Department began publishing a newsletter called *First Element* to report on progress of the various hydrogen research projects it continues to fund.[34] A total of $3.2 million in provincial funds had been invested in hydrogen research in Alberta through the end of 1993.[35]

British Columbia

The British Columbia Ministry of Energy and the provincial bus operator, BC Transit, are participants in the development of a commercial fuel cell bus by Vancouver-based Ballard Power Systems.[36] The province of British Columbia has provided financial support for the Ballard Bus project, discussed below. The Institute of Integrated Energy Systems at the University of Victoria is a second center of hydrogen research activity in British Columbia. The Institute has published a number of energy policy studies and analyses of specific proposals to use hydrogen in niche markets, such as for emergency telephone power supplies.[37] In 1995, the Institute initiated a five-year fuel cell research program called the "Next Generation of Fuel Cells for Transportation." This $5 million effort is funded by the university, the natural gas company British Gas, the Natural Sciences and Engineering Research Council of Canada, and Ballard Power Systems.[38]

The Ballard Bus Project: World's First Hydrogen Fuel Cell Bus

On June 6, 1993, the world's first bus powered solely by a hydrogen fuel cell was driven down the streets of Vancouver, British Columbia. Built by Ballard Power Systems and operated as part of the BC Transit fleet, the bus is a prototype version of a fuel cell bus that the company hopes to sell commercially beginning in 1998.[39]

Ballard Power Systems was formed in 1979 to develop advanced propulsion systems, including fuel cells and batteries, for specialized applications such as submarines. In 1984 Ballard turned its attention to proton-exchange membrane fuel cell technology. In November 1990 it began designing its first fuel-cell-powered bus. Ballard employs about 150 people at its manufacturing site in North Vancouver.[40]

The Ballard proton-exchange membrane fuel cell bus

- Prototype bus unveiled in mid-1993 is based on a conventional 32-foot bus built by National Coach Corporation in California; it carries up to 20 passengers.
- Diesel engine has been replaced by 24 stacks of proton-exchange membrane fuel cells each with a 5-kilowatt capacity. Total output of 120 kilowatts is roughly equivalent to a conventional 160-horsepower diesel engine
- At 24,000 pounds, the Ballard bus is only 1,000 to 3,000 pounds heavier than the conventional bus. The fuel cell stack itself weighs less than 1,000 pounds.
- Hydrogen is carried on board in compressed form. The nine compressed gas storage cylinders provide enough hydrogen for a 100-mile driving range.
- Acceleration from zero to 30 miles per hour takes about 20 seconds.[41]

While Ballard was the overall project director and developer of the fuel cell, Science Applications International Corporation of San Diego provided propulsion, motor, and other systems integration engineering for the bus, and Mohawk Oil of Vancouver provided the hydrogen. Since its initial run, the Ballard bus has been shipped to several locations for demonstration purposes. In August 1993, for example, it toured California.[42] In August 1994, the Ballard fuel cell bus was used throughout the weeklong Commonwealth Games in Vancouver to carry spectators, including Queen Elizabeth and Princes Philip and Edward, between various sporting events.[43]

In Phase I, the project developed the prototype Ballard bus over a four-year period at a cost of about $6 million, obtained mostly from Ballard and the national and British Columbia governments. Phase II of the Ballard Bus Project, initiated in 1994, includes development and commercialization of a full-size urban bus, which is expected to take an additional two to three years and to cost an additional $6 million. Funding for the Phase II commercial bus development has been secured, including funds from Ballard itself, which has pledged to contribute about $1.5 million. The

Ballard Power Systems plans to commercialize production of its fuel cell bus. Photo: Hydrogen Consultants, Inc.

bulk of funds came from the British Columbia government ($2.2 million), the California South Coast Air Quality Management District ($1.3 million), and the Canadian national government ($1 million).[44]

The commercial version of the Ballard bus will be a modified standard 40-foot urban transit bus. A battery storage system equipped with regenerative braking will be added to the hydrogen fuel-cell system to increase efficiency and power. An advanced fuel cell with nearly twice the power as the prototype version will be installed. The total peak power output of this hybrid bus will be about 205 kilowatts. Improved efficiencies and expanded hydrogen storage will provide a driving range of 350 miles and a maximum passenger load of 60 people.[45]

In late 1994, the Canadian bus manufacturer, New Flyer Industries, Ltd., joined the Phase II bus development effort. New Flyer will provide the bus – a 40LF, low-floor model into which the fuel cell propulsion system will be installed. The companies admit that the early buses will cost twice as much as conventional diesel-powered buses, but they say that larger production volumes will permit competitive pricing and much improved environmental performance compared with diesel buses.[46]

In addition to its bus project, Ballard is also developing its proton-exchange membrane fuel cell technology for other applications. As discussed above, Ballard has joined with Daimler-Benz of Germany to develop a fuel cell for use in an automobile. Ballard has also shipped fuel cells for testing by other automotive manufacturers, including Peugeot and Renault in France.[47] In mid-1993, Ballard signed an agreement with Dow Chemical Company to develop a fuel cell powered by reformed natural gas for use in stationary applications to generate electricity. Dow Chemical manufactures membranes that can be used in proton-exchange membrane fuel cells.[48]

Multinational Hydrogen Projects

Several international hydrogen projects that link energy-producing regions with countries with high energy consumption are currently underway. Initiatives to construct solar-hydrogen production and use facilities include the Euro-Quebec project, the Hysolar project, the Eureka Bus project, the HYPASSE project, the "Norwegian Hydro Energy in Germany" project, and the Great Sahara Project.

Table 2. Major International Solar Hydrogen Production Facilities

Solar-Wasserstoff-Bayern Project (Germany)
Participants: The private companies Bayernwerk, Linde, Siemens, and MBB and German government agencies.
270 kilowatts from photovoltaic cells power an electrolysis unit that separates hydrogen from water. Facility includes a liquefied hydrogen refueling station (1986-present).

Euro-Quebec Project (Canada and the European Union)
Participants: Quebec government, European Union, and 44 industrial partners.
Proposes to use 100 megawatts of electricity generated at a hydroelectric plant to produce hydrogen from water. Plans to transport hydrogen to Europe. Demonstration projects involving buses in Europe and Montreal are underway. (development 1989-present).

Hysolar Project (Germany and Saudi Arabia)
Participants: German and Saudi Arabian governments and universities.
362 kilowatts of electricity from three photovoltaic-powered electrolyzers used to produce hydrogen. Includes demonstration in stationary and vehicle applications (1989-present).

The Euro-Quebec Project

Euro-Quebec Hydro-Hydrogen Pilot Project, a joint undertaking of governments and industries in Europe and Canada, proposes to dedicate the electricity generated at a 100-megawatt hydroelectric plant in Canada for use in producing hydrogen from the electrolysis of water. The hydrogen will be shipped to energy markets in Europe.[49] The project was initiated in 1989 with a treaty between the government of Quebec and the Commission of the European Communities, now known as the European Union, represented by the Joint Research Institute in Ispra, Italy. As of the beginning of 1994, the project included 44 industrial partners. A Joint Management Group of two companies directs the project: Ludwig-Bolkow-Systemtechnik in Germany and Hydro-Quebec Company in Canada.

The project has designated a power plant at Sept-Iles on the St. Lawrence River in Quebec as the site for this effort. The world's largest electrolyzers will be built at the power plant to produce up to 24,000 cubic meters of hydrogen per hour. The gaseous hydrogen will probably be liquefied to permit economical transportation in tankers across the ocean.[50]

The proposed European delivery point for the hydrogen is Hamburg, Germany. Once unloaded, the hydrogen will be tested in a variety of end-use applications in Hamburg and elsewhere in Europe. The current proposed applications include city buses, an Airbus airplane, two advanced gas-fired power generators, and one fuel-cell-equipped generator. A portion of the hydrogen could also be blended with natural gas and shipped through existing pipelines to serve natural gas markets. Calculations of the energetics of the entire fuel-cycle indicate that about 55 percent of the 100 megawatts of electrical output in Canada would be delivered as chemical energy in hydrogen at the end uses in Europe.

Spending on the Euro-Quebec Project from its inception in 1987 through the end of 1993 totalled about $59 million, with several million more allocated for the demonstration phase of the project from 1994 through 1997. Of this amount, the governments of the European Union have paid about 45 percent, European industries paid another 36 percent, and Canadian industries and the Quebec government split the remaining 19 percent. The capital cost of implementing the full-scale Euro-Quebec Project is estimated to exceed $500 million. Project participants have not committed funds to complete the project past the three phases of planning and component demonstration. Moreover, progress in the 1990s has slowed as concern has mounted about the potential adverse environmental effects of the hydroelectric power development needed in Quebec to produce the hydrogen.[51]

Phases I and II of the Euro-Quebec Project were completed between 1987 and 1991. They included initial project concept development, completion of the treaty officially establishing the project, and a series of technical studies defining the project components in detail. Phase III of the project, through 1998, will include further technical and economic analysis of the hydrogen production and shipment scheme, plus development of more specific programs for the end-use demonstration of the hydrogen in Europe.

Four European hydrogen bus projects – one each in Germany, Belgium, Italy, and Ireland – and one bus project in Canada are part of the Euro-Quebec Project's Phase III demonstration program. The average cost of each bus project will be about $3 million.[52]

Germany

The German effort, called the Bavarian Bus Project, includes the modification of a SL 202 model city bus, manufactured by the German bus builder MAN, to burn liquefied hydrogen. Early testing and refueling of the bus will take place at the existing hydrogen test facility at BMW headquarters in Munich. The first bus is scheduled for demonstration in 1995. Several additional hydrogen buses, including a shuttle bus to the Munich airport, will operate during the next stage of the project, followed by a demonstration of a fleet of up to 10 hydrogen buses later this decade. Sponsors of the Bavarian Bus Project include MAN, BMW, Linde (which will supply the hydrogen), the Bavarian State Ministry for State Development and Environmental Affairs, and the Ludwig-Bolkow-Systemtechnik GmbH, which will provide overall project management.[53]

Belgium

Sponsored by Hydrogen Systems NV, this project will develop and test an internal combustion engine-powered bus fueled by liquefied hydrogen. Nicknamed the "Greenbus," it began field testing in Brussels and Ghent in late 1994. In 1993, Hydrogen Systems achieved another important milestone by completing its first hydrogen vehicle, a dual-fuel minivan equipped with a metal hydride storage system for hydrogen.[54]

Italy

This demonstration involves construction of a liquefied hydrogen-powered fuel-cell bus by a team led by Ansaldo Ricerche. Using a fuel cell built by De Nora Permelec, the bus will be tested in the northern Italian city of Brescia.[55]

Ireland

The Irish project will include a mid-size hybrid electric bus equipped with a hydrogen-powered combustion engine that will generate electricity to propel the bus. The engine will be manufactured by Kockumas Marine, and the bus will be tested in Dublin.

Canada

The Canadian project involves the conversion of three buses in Montreal to run on Hythane, a mixture of natural gas and hydrogen. As discussed earlier in this chapter, two of these buses are already operating on Montreal streets.

The Hysolar Project

The Hysolar project was initiated in the late 1980s as a joint undertaking with equal financial participation by Germany and Saudi Arabia. The project aims to produce hydrogen with solar photovoltaic cells and to utilize this fuel in stationary and vehicle applications. The German participants include the BMFT, the DLR, the University of Stuttgart, and the State of Baden-Wuerttemberg, which includes Stuttgart. Participants from Saudi Arabia include the King Abdulaziz City of Science and Technology and the universities of Dhahran, Jeddah, and Riyadh.[56]

The two-phase project will cost over $60 million by the time it is completed in 1996. Phase I, completed in 1993, involved the construction of three solar-hydrogen production facilities with a total electrical-generating capacity of 362 kilowatts, two in Saudi Arabia and one in Germany, at a total cost of about $37 million. Since 1986, more than 100 professional papers have been published summarizing the results of various components of Phase I of the Hysolar project, marking it as one of the largest centers of hydrogen research and development activities in the world today. Annual expenditures in 1994 were about $4.7 million, split equally between German and Saudi interests.

The three demonstrations each involve arrays of solar photovoltaic cells. The two demonstration facilities in Saudi Arabia include a 350-kilowatt pilot plant as part of the existing "Solar Village" in Riyadh and a 2-kilowatt laboratory facility in Jeddah. The Riyadh pilot plant is capable of producing 170,00 cubic meters of hydrogen each year, ranking it among the largest solar-hydrogen production facilities currently operating in the world. The German Hysolar facility, operational since 1989, is located on the University of Stuttgart campus. The original 10-kilowatt photovoltaic generating facility was expanded in 1990 to include a comparably sized solar thermal generating system. The Stuttgart plant includes 720 individual solar photovoltaic modules and one large solar concentrator. The electricity, once generated at the three Hysolar demonstration facilities, is used by different types of advanced electrolysis technologies to generate hydrogen.

Phase I was originally to be completed in 1991, but a series of problems at the 350-kilowatt Riyadh facility delayed the official dedication until February 6, 1994.

Phase II, a four-year effort which began in 1992 concurrent with the final stages of Phase I, includes the testing of technologies, such as fuel cells and catalytic burners, to utilize the hydrogen produced at the solar-hydrogen demonstration facilities. Construction and testing of at least one hydrogen fuel-cell-powered vehicle is included in Phase II of the Hysolar project.

Other Cooperative Efforts

Several smaller multinational hydrogen projects are also under development.

The Eureka Bus Project

This joint venture involves companies from four European countries that seek to commercialize a hydrogen-powered bus. Fueled by liquefied hydrogen, the full-size articulated urban bus – which has been under development since 1989 – is powered by a 78-kilowatt alkaline fuel cell. Nickel-cadmium batteries provide auxiliary power. The lead company is Elenco NV, a fuel cell developer headquartered in Belgium. Other participants include the Netherlands affiliate of the hydrogen producer Air Products and Chemicals, the Italian energy company Ansaldo, and Saft, a French battery manufacturer. The joint venture unveiled the bus at a press conference in Brussels in late 1994. Public transit authorities in Brussels and Amsterdam will test the Eureka bus starting in 1995.[57]

The HYPASSE Project

A joint project between Germany and Switzerland, called HYPASSE, seeks to use hydroelectricity generated in Switzerland to produce hydrogen for consumption in German energy markets. The key sponsors of HYPASSE, which stands for "**Hy**drogen-**P**owered **A**pplications Using **S**easonal and Weekly **S**urpluses of **E**lectricity," are Daimler-Benz in Germany and the Paul Scherrer Institute, a government-sponsored research organization in Switzerland. The project expects to use surplus electrical generation capacity in Switzerland, mostly during nights and weekends, to electrolyze water – producing hydrogen. The hydrogen would be trucked or shipped by pipeline to Germany. The proposed initial German market for the hydrogen would be to power urban buses built by Daimler-Benz as a result of its bus project discussed earlier in this chapter. The first planning phase of this project, through the end of 1995, will cost approximately $17.3 million.[58]

The Norwegian Hydro-Energy in Germany Project

The Norwegian project, similar in concept to the Euro-Quebec Project, would use electricity generated at the Glomfjord hydropower plant in Norway to produce hydrogen by electrolysis of water. The hydrogen would be liquefied and shipped by boat to Germany in 40-foot containers. Once in Germany, the hydrogen could be used in any of the end-use applications, including transportation, envisioned in the Euro-Quebec project. German involvement in the NHEG is coordinated by Ludwig-Bolkow-Systemtechnik in Munich. Norwegian participation is led by the utility company Norsk Hydro.[59]

The Great Sahara Project

Located in Northern Africa, this is another project that aims to provide hydrogen to Europe. Sponsored by the International Energy Foundation in Tripoli, and other organizations in Libya, the Great Sahara Project proposal includes a large solar thermal power plant to be built near the desert community of Ghadames, about 300 miles southwest of Tripoli. Hydrogen produced by electrolysis could then be liquefied and shipped to Europe by tanker. Alternatively, the hydrogen could be mixed with Algerian natural gas and transported to Europe through an existing pipeline under the Mediterranean Sea to Italy.[60]

International Coordination of Hydrogen Energy Programs

Over 30 nations currently have some research and development activities underway related to hydrogen energy. A number of organizations have formed or are forming to coordinate these activities.

- **International Association of Hydrogen Energy** (Coral Gables, Florida) Since 1974, the International Association of Hydrogen Energy has been the principal professional organization representing government, academic, and private companies active in hydrogen energy. The association "strives to advance the day when hydrogen energy will become the principal means by which the world will achieve its long-sought goal of abundant clean energy for mankind."[61]

  The monthly *International Journal of Hydrogen Energy* is the official publication of the Association and its major work product. Its other major activity is to convene biannual conferences. The first World Hydrogen Conference was held in Florida in 1976. In 1992, more than 300 delegates from 35 countries attended the Ninth World Hydrogen Conference in Paris. The 10th Conference drew nearly 500 people to Cocoa Beach, Florida in June 1994. Over 160 professional papers were presented at the 1994 conference and published in the conference proceedings.[62]

- **The International Energy Agency** (Paris) Part of the Organization of Economic Cooperation and Development, the International Energy Agency is active in hydrogen energy development on behalf of its 23 member countries. In 1977, 10 of these countries formed the Hydrogen Implementing Agreement to collect and share information about hydrogen development efforts around the world. Twelve committees, called Annexes, have been formed since then to focus on various aspects of hydrogen production and use.[63] In December 1994, the IEA published a review of hydrogen programs currently underway in 11 countries.[64]

At least two additional international organizations are in the process of forming:

- **United Nations Center** (Istanbul) In 1992, the United Nations Industrial Development Organization announced a program to establish an international center for hydrogen energy and technology in Istanbul. The center would include a research center where emerging hydrogen technologies could be tested. An objective of the center would be to facilitate the use of hydrogen energy by developing nations. Project sponsors are seeking support for the center from member countries in the United Nations.[65]

- **International Council on Cooperation and Development of Hydrogen** (Montreal) Another international organization is in the formation stage under the sponsorship of the Canadian Hydrogen Industry Council. In November 1993 the council convened in Montreal the "First World Hydrogen Summit," attended by 46 government and industry officials. The purpose of the meeting was to discuss the establishment of an International Council on Cooperation and Development of Hydrogen to complement the hydrogen programs of existing international agencies. Funded by an initial $50,000 grant from the Montreal International Conference Centre Corporation, project sponsors in 1994 are establishing a constituting board to guide the formation of the council.[66]

Notes

1. International Energy Agency, *Hydrogen Energy Activities in Eleven IEA Countries* (International Planning Associates Inc., Silver Spring, MD, December 1994), p. 19.
2. Information contained in this section was obtained largely from interviews conducted during three field trips to Europe, in 1991, 1992, and 1993. Other specific sources are cited below.
3. Peter Hoffman, *The Forever Fuel: The Story of Hydrogen* (Westview Press, Boulder, Colorado, 1981), pp. 25-33.
4. Most of the background information about German government activities in hydrogen energy was obtained during a field visit to the German DLR offices in Stuttgart, Germany, in May 1991.
5. German Federal Ministry for Research and Technology, *Alternative Energy Sources for Road Transport: Hydrogen Test Drive* (German Federal Ministry for Research and Technology, Bonn, Germany, 1990).
6. Most of the background information about Daimler-Benz hydrogen vehicle programs was obtained during a field visit to the Daimler-Benz research headquarters in Stuttgart, Germany, in May 1991.
7. International Energy Agency, *Hydrogen Energy Activities in Eleven IEA Countries, op. cit.*, p. 19.
8. "Daimler-Benz Sees Hydrogen Applications in City Buses," *Clean Fuels Report*, September 1992.
9. "Hydrogen to Power Munich Airport Vehicles," *The Hydrogen Letter*, November 1994.
10. "Modified Diesel for Planned German H Bus Line," *Hydrogen & Fuel Cell Letter*, February 1995.
11. "Ballard, Daimler-Benz Launch Fuel Cell Program," *The Hydrogen Letter*, June 1993.
12. "Daimler Unveils Electric Vehicle Using Fuel Cells," *The Wall Street Journal*, April 14, 1994; and "Daimler Unveils FC Van," *The Hydrogen Letter*, May 1994.

13. Most of the background information about BMW hydrogen vehicle programs was gathered during a field visit to the BMW research headquarters in Munich, Germany, in May 1991.

14. D. Reister and W. Strobl, "Current Development and Outlook for the Hydrogen-Fueled Car," *Hydrogen Progress IX: Proceedings of the 9th World Hydrogen Conference* (Paris, France, 1992).

15. "BMW Plans '95 LH_2 Car Tour for California, Canada," *The Hydrogen Letter*, August 1994.

16. "Demonstration Plant Neunburg vorm Wald/Germany to Investigate and Test Solar-Hydrogen Technology," a paper published by Solar-Wasserstoff-Bayern GmbH in 1992; J. Tachtler and A. Szyska, "Car Fueling with Liquid Hydrogen, Neunburg vorm Wald Solar Hydrogen Project: Experience and Results of First Project Phase, Concept for Second Phase," *International Journal of Hydrogen Energy*, Vol. 19, No. 4, April 1994; and A. Szyska, "Neunburg vorm Wald - Test Centre for Solar Hydrogen Technology," *International Journal of Hydrogen Energy,* Vol. 19, No. 10, October 1994.

17. "German Renewables Funding in Cut Back," *The Hydrogen Letter*, February 1993 and "German Government Lukewarm Towards Hydrogen," *The Hydrogen Letter*, October 1993.

18. International Energy Agency, *Hydrogen Energy Activities in Eleven IEA Countries, op. cit.,* p. 6.

19. Background information about the Japanese hydrogen program was obtained largely during a field trip to Japan in June 1993. Additional background sources include International Energy Agency, *Hydrogen Energy Activities in Eleven IEA Countries, op. cit.,* p. 27-33.

20. International Energy Agency, *Hydrogen Energy Activities in Eleven IEA Countries, op. cit.,* p. 6.

21. "Hydrogen Technology Demonstrated at Musashi Institute of Technology," *Clean Fuels Report,* April 1992; "Liquid Hydrogen Truck with Icebox Demonstrated," *Clean Fuels Report,* September 1993; "The Power System of a Computer Controlled Hydrogen Car – GH_2 Injection and Spark Ignition Engine with LH_2 Tank and Pump," reprint of a professional paper published in 1991 by S. Furuhama and others at the Musashi Institute of Technology; and "LH_2 Application to a Hydrogen Fueled Truck with an Icebox on Board," *Proceedings of the New Energy Systems and Conversions Conference* (Yokohama, Japan, June 27-30, 1993).

22. "Mazda Takes Journalists for a Jaunt in Hydrogen-Powered Automobile," *The Wall Street Journal,* October 25, 1993; "Mazda Plans Hands-On Hydrogen Show Cars," *Clean Fuels Report,* August 1993; "Hydrogen Miata: A Step Toward Fleet Tests," *The Hydrogen Letter,* August 1993; "Mazda Unveils New Hydrogen Prototype," *The Hydrogen Letter,* November 1993; "Mazda, Electrolyzer Share Hydrogen Displays," *The Hydrogen Letter,* March 1994; "More Clean Fuel Vehicles Were Introduced at Tokyo Auto Show," *Clean Fuels Report,* February 1992; and "Mazda Develops Hydrogen Fueled Rotary Engine," *Clean Fuels Report,* April 1992.

23. C. Watanabe and S. Katayama, "Japan's New Sunshine Program and International Clean Energy Network Using Hydrogen Energy," *Proceedings of the 4th National Hydrogen Conference (*Washington, DC, March 1993); and the following articles from *The Hydrogen Letter*, "Japan Confirms $2 Billion, 28-Year Hydrogen Plan," April 1993, "Japan Gears Up for WE-NET Hydrogen Project," August 1993, and "Four Foreign Organizations Get WE-NET Contracts," January 1995; and "Japan's WE-NET Hydrogen Program Takes Shape," *Clean Fuels Report*, November 1993.

24. "Japan Presents Trans-Asia Pipeline Concept," *Hydrogen & Fuel Cell Letter*, March 1995.

25. Hydrogen Industry Council, "A Strategic Choice for Your Company" (Hydrogen Industry Council, Montreal, Quebec, Canada), not dated.

26. Advisory Group on Hydrogen Opportunities, *Hydrogen: National Mission for Canada* (Advisory Group on Hydrogen Opportunities, Ottawa, Canada, June 1987).

27. The Hydrogen Industry Council, *Report on Activities 1988* (The Hydrogen Industry Council, Montreal, Canada, 1989).

28. The Hydrogen Industry Council, "A Strategic Choice...," *op. cit.*

29. "Canadian Government to Fund Hydrogen Research," *Clean Fuels Report*, June 1991; and "Hydrogen Energy Activities in Canada Focus on Technical Objectives," *Clean Fuels Report,* June 1995.

30. International Energy Agency, *Hydrogen Energy Activities in Eleven IEA Countries, op. cit.,* pp. 10-17.

31. R. Greven, "Hydrogen-Related Projects in Ontario," *Proceedings of the 4th National Hydrogen Conference* (Washington, DC, March 25, 1993).

32. "Quebec's Hythane Bus to Start Test Runs," *The Hydrogen Letter*, January 1995; and "Hythane Bus Unveiled in Canada," *Clean Fuels Report*, February 1995.

33. Alberta Energy Department, *Some Studies of Hydrogen Technologies Suitable for Alberta* (Alberta Energy Department, Edmonton, Canada, 1992).

34. Alberta Energy Department, *First Element* (Alberta Energy Department, Edmonton, Canada, January 1994).

35. Alberta Energy Department, *Alberta Hydrogen Energy Program: Annual Review 1991/1992* (Edmonton, Canada, 1993).

36. "Ballard to Build ZEV Transit Bus," *Fuel Cell News,* Winter 1994.

37. J. Wells and D. Scott, "Fuel Cells for Telephone Networks," *International Journal of Hydrogen Energy*, May 1994.

38. "IESVic Starts 'Next Generation' Fuel Cell Program," *The Hydrogen Letter*, January 1995.

39. Ballard Power Systems, press release (Vancouver, Canada, June 8, 1993).

40. "Company Moves Hydrogen Fuel Dream Closer to Reality," *Riverside Press Enterprise* (Riverside, CA. February 8, 1992); and presentation by Paul Howard of Ballard Power Systems to the Western Interstate Energy Board (Victoria, Canada, June 4, 1993).

41. Paul Howard presentation, *op. cit.*; and "First Hydrogen Fuel Cell Powered Bus," *Hydrogen Today*, Vol. 5, No. 1, Winter 1994.

42. Ballard Power Systems, press release, (Vancouver, Canada, August 31, 1993).

43. "Ballard Bus Works Out at Commonwealth Games," *The Hydrogen Letter*, September 1994.

44. "Canada and California Will Provide New Funding for Ballard Fuel-Cell Bus," *Clean Fuels Report*, April 1994.

45. Paul Howard presentation, *op. cit.*; and "Ballard Plans Commercial Fuel Cell Bus Prototype," *The Hydrogen Letter*, March 1994.

46. "Ballard, New Flyer, to Market Fuel Cell Buses," *The Hydrogen Letter*, December 1994.

47. Ballard Power Systems, press release, (Vancouver, Canada, September 7, 1993).

48. *Dow News,* press release, July 8, 1993; and "Powerful Solutions In a Changing Marketplace: Proton Exchange Membrane Fuel Cell Technology," Dow Chemical Co., 1994.

49. J. Gretz *et al.,* "Status of the Hydro-Hydrogen Pilot Project," *International Journal of Hydrogen Energy*, Vol. 19 No. 2, February 1994, a paper with the same title presented by R. Wurster, *Proceedings of the 4th National Hydrogen Conference* (Washington, DC, March 1993); and "It's Official: EQHHPP Announces Demo Projects," *The Hydrogen Letter,* November 1993.

50. "Euro-Quebec Project Finds Liquid Hydrogen Most Economical Route," *Clean Fuels Report,* June 1991.

51. "Greens Attack Euro-Quebec Project, Miss Target," *The Hydrogen Letter*, May 1993.

52. "It's Official: EQHHPP Announces Demo Project," *The Hydrogen Letter*, November 1993; "Euro-Quebec Hydro-Hydrogen Pilot Project Summarized," *Clean Fuels Report*, April 1994; and "Status of Euro-Quebec Hydro-Hydrogen Pilot Project Updated," *Clean Fuels Report*, September 1994.

53. R. Wurster *et al.,* "Application of LH_2 Cars and Urban Buses in Munich," *Hydrogen Progress IX: Proceedings of the 9th World Hydrogen Conference* (Paris, France, 1992).

54. Hugo Vandenborre, Hydrogen Systems NV, *Proceedings of the 4th National Hydrogen Conference* (Washington, DC, March 25, 1993).

55. "Italian Hydrogen Research Program Emphasizes Vehicle Applications," *Clean Fuels Report*, June 1995.

56. HYSOLAR, "Hysolar: A German-Saudi Arabian Partnership" (Stuttgart, Germany 1989); A. Brinner *et al.,* "Test Results of the Hysolar 10 kW PV-Electrolysis Facility," *International Journal of Hydrogen Energy*, Vol. 17, No. 3, March 1992; H. Steeb *et al.,* "Hysolar: An Overview on the German-Saudi Arabian Program on Solar Energy," *Hydrogen Progress IX: Proceedings of the 9th World Hydrogen Conference* (Paris, France, June 1992); and "Saudi-German HYSOLAR Plant Starts Operation," *The Hydrogen Letter*, March 1994.

57. P. Hoogeveen, "Eureka Fuel Cell Bus," *Hydrogen Progress IX: Proceedings of the 9th World Hydrogen Conference* (Paris, France, June 1992); and Eureka Project, information package, received by INFORM, July 1992; and "Belgium's Eureka Fuel Cell Bus Takes to Road," *The Hydrogen Letter*, December 1994.

58. "WHEC-10: HYPASSE," *The Hydrogen Letter*, August 1994; and J. Zieger, "HYPASSE – Hydrogen Powered Automobiles Using Seasonal and Weekly Surplus of Electricity," *Hydrogen Energy Progress X: Proceedings of the 10th World Hydrogen Energy Conference* (Cocoa Beach, FL, June 20-24, 1994).

59. K. Andreassen *et al.,* "Norwegian Hydro Energy in Germany," *International Journal of Hydrogen Energy*, Vol. 18, No. 4, April 1993.

60. G. Elrushi and J. Zubia, "Solar Hydrogen: The Great Sahara Project," *International Journal of Hydrogen Energy*, Vol. 19, No. 3, March 1994.

61. "Foreword," *International Journal of Hydrogen Energy*, Vol. 20, No. 7, July 1995. Each issue of this *Journal* includes a description of the International Association for Hydrogen Energy.

62. T. Veziroğlu, "Twenty Years of the Hydrogen Movement," *International Journal of Hydrogen Energy*, Vol. 20, No. 1, January 1995.

63. "IEA Groups Discusses New Strategies, Approaches," *The Hydrogen Letter*, April 1993; and "US, EU, Japan, Are New Co-Chairs of IEA Hydrogen Executive Committee," *Hydrogen and Fuel Cell Letter*, June 1995.

64. International Energy Agency, *Hydrogen Energy Activities in Eleven IEA Countries, op. cit.*

65. "UNIDO Plans Hydrogen Center in Turkey," *The Hydrogen Letter*, June 1993.

66. "International Hydrogen Industry Group Formed," *The Hydrogen Letter*, December 1993.

Bibliography

The following documents are the major publications that INFORM has reviewed during alternative fuel projects addressing natural gas and hydrogen vehicle technology.

Advisory Group on Hydrogen Opportunities. *Hydrogen: A National Mission for Canada*. Ottawa, Canada: Ministry of State for Science and Technology, June 1987.

Allison Gas Turbine Division, General Motors Corp. *Research and Development of Proton-Exchange Memrane (PEM) Fuel Cell System for Transportation Application*. Prepared for the US Department of Energy, Office of Transportation Technologies, Washington, DC: November 30, 1993.

American Automobile Manufacturers Association. *Facts & Figures '93*. Detroit, Michigan: 1993.

American Automobile Manufacturers Association. *World Motor Vehicle Data: 1993 Edition*. Detroit, Michigan: 1993.

American Gas Association. "An Analysis of the Economic and Environmental Effects of Natural Gas as an Alternative Fuel." Arlington, Virginia: December 15, 1989.

American Gas Association. *1993 Gas Facts*. Arlington, Virginia: 1994.

Austin, Thomas *et al., Potential Emissions and Air Quality Effects of Alternative Fuels*. Sacramento, California: Sierra Research, Inc., March 28, 1989.

Billings, Roger E. *The Hydrogen World View*. Independence, Missouri: American Academy of Science, 1991.

Bockris, J. O'M. *Energy Options: Real Economics and the Solar Hydrogen System*. Sydney, Australia: ANZ Book Co., 1980.

Bockris, J. O'M, Nejat Veziroğlu, and Debbi Smith. *Solar Hydrogen Energy: The Power to Save the Earth*. London: MacDonald & Co., 1991.

Braun, Harry. *The Phoenix Project: An Energy Transition to Renewable Resources*. Phoenix, Arizona: Research Analysts, 1990.

California Council for Environmental & Economic Balance. *Alternative Fuels as an Air Quality Improvement Strategy*. Sacramento, California: November 1987.

California Energy Commission. *AB 234 Report: Cost and Availability of Low-Emission Motor Vehicle Fuels*. Sacramento, California: August 1989. *AB 234 Report Update*, August 1991.

Clean Energy Research Institute. *Solar Hydrogen Energy System*. (Prepared under contract to the US Department of Energy, Solar Energy Research Institute.) Coral Gables, Florida: October 1989.

Clinton, William and Albert Gore. *The Climate Change Action Plan*. Washington, DC: The White House, October 1993.

Congress of the United States, Office of Technology Assessment. *Changing by Degrees: Steps to Reduce Greenhouse Gases*. Washington, DC: US Congress. 1991.

Congress of the United States, Office of Technology Assessment. *Replacing Gasoline: Alternative Fuels for Light-Duty Vehicles*. Washington, DC: US Congress. 1990.

Congress of the United States, Office of Technology Assessment. *Saving Energy in US Transportation*, Washington, DC: US Congress. July 1994.

Davis, S. *Transportation Energy Data Book: Edition 14*. Oak Ridge National Laboratory, Oak Ridge, Tennessee: ORNL-6798, May 1994.

DeLuchi, Mark. *Emissions of Greenhouse Gases from the Use of Transportation Fuels and Electricity*. Argonne, Illinois: Argonne National Laboratories, 1992.

DeLuchi, Mark. *Hydrogen Fuel-Cell Vehicles*. Davis, California: Institute of Transportation Studies, September 1992.

DeLuchi, Mark *et al.,* "Methanol vs. Natural Gas Vehicles: A Comparison of Resource Supply, Performance, Emissions, Fuel Storage, Safety, Costs, and Transitions." Warrendale, Pennsylvania: Society of Automotive Engineers, 1988.

DeLuchi, Mark and Joan Ogden. "Solar-Hydrogen Fuel-Cell Vehicles," *Transportation Research-A*, Vol. 27A #3, 1993.

Ebasco Services Inc. *Hazard Assessment of Natural Gas Vehicles in Public Parking Garages*. New York, New York: July 1991.

Electric Power Research Institute. *Proceedings: Annual Meeting of the National Hydrogen Association*. Palo Alto, California: Published annually since 1990.

Energy International Inc. *Light Duty Vehicle Full Fuel Cycle Emissions Analysis*. Chicago, Illinois: Gas Research Institute, April 1994.

Environmental and Business Groups Working on Sustainable Energy Policies. "Twenty Years After the Oil Embargo: A Review of Energy Trends Since the 1973 Oil Embargo and a Forecast for the Next Two Decades." Washington, DC: October 1993.

Federal Fleet Conversion Task Force. *First Interim Report: August 1993*. (DOE/PO-0001) Washington, DC: 1993.

Federal Highway Administration. *Highway Statistics 1992*. Washington, DC: Government Printing Office # FHWA-PL-93-023, 1993.

Fisher, Diane. *Reducing Greenhouse Gas Emissions With Alternative Transportation Fuels*. Oakland, California: Environmental Defense Fund, April 1991.

Gas Research Institute. *A White Paper: Preliminary Assessment of LNG Vehicle Technology, Economics, and Safety Issues*. Chicago, Illinois: Gas Research Institute, January 1992.

Gas Research Institute. *1994 Edition of the GRI Baseline Projection of US Energy Supply and Demand to 2010*. Chicago, Illinois: Gas Research Institute, June 1994.

German Federal Ministry for Research and Technology. *Alternative Energy Sources for Road Transport: Hydrogen Test Drive*. Bonn, Germany: Verlag TUV Rheinland GmbH, 1990.

German Federal Ministry for Research and Technology. "Summary of Report on Solar Hydrogen Energy Economy." Bonn, Germany: April 1988.

Gordon, Deborah. *Steering a New Course: Transportation, Energy, and the Environment*. Cambridge, Massachusetts: Union of Concerned Scientists, 1991.

Gushee, David. *Impact of Highway Fuel Taxes on Alternative Fuel Vehicle Economics* (March 16, 1994) and *Disparate Impacts of Federal and State Highway Taxes on Alternative Motor Fuels* (December 17, 1993). Washington, DC: Congressional Research Service.

Hoffman, Peter. *The Forever Fuel: The Story of Hydrogen*. Boulder, Colorado: Westview Press, 1981.

Hwang, Roland *et al., Driving Out Pollution: The Benefits of Electric Vehicles*. Berkeley, California: Union of Concerned Scientists, Revised Edition, November 1994.

Interagency Commission on Alternative Motor Fuels. *First Interim Report of the Interagency Commission on Alternative Motor Fuels.* Washington, DC: September 30, 1990. Second Interim Report, September 1991.

International Association for Hydrogen Energy. *Hydrogen Energy Progress: Proceedings of the 8th World Hydrogen Energy Conference,* Honolulu and Waikoloa, Hawaii, July 22-27, 1990. Elmsford, New York: Pergamon Press, 1990.

International Association for Hydrogen Energy. *Hydrogen Energy Progress: Proceedings of the 9th World Hydrogen Energy Conference*, Paris, France, June 22-25, 1992. Elmsford, New York: Pergamon Press, 1992.

International Association for Natural Gas Vehicles. *A Position Paper on Natural Gas Vehicles – 1993*. Auckland, New Zealand: IANGV, 1993.

International Association for Natural Gas Vehicles. *Conference Proceedings NGV90: 2nd International Conference and Exhibition*, Buenos Aires, Argentina, October 21-25, 1990. Auckland, New Zealand: IANGV, 1990.

International Association for Natural Gas Vehicles. *Conference Proceedings NGV92: 3rd Biennial International Conference and Exhibition,* Goteborg, Sweden, September 22-25, 1992. Auckland, New Zealand: IANGV, 1992.

International Energy Agency. *Cars and Climate Change*. Paris, France: 1993.

International Energy Agency. *Greenhouse Gas Emissions: The Energy Dimension*. Paris, France: 1991.

International Energy Agency. *Methane as a Motor Fuel.* (Prepared by Sypher: Mueller International Inc.) Paris, France: May 1992.

International Energy Agency. *Substitute Fuels for Road Transport.* Paris, France: 1990.

International Gas Union and the International Association of Natural Gas Vehicles. *Task Force Report: Milan 1994*. England: Greenaways, 1994.

Kiernan, Patrick. *Hydrogen: The Invisible Fuel*. Aspen, Colorado: IRT Environment of Aspen, April 1991.

MacKenzie, James. *The Keys to the Car*. Washington, DC: World Resources Institute, 1994.

MacKenzie, James and Michael Walsh. *Driving Forces: Motor Vehicle Trends and Their Implications for Global Warming, Energy Strategies, and Transportation Planning*. Washington, DC: World Resources Institute, 1990.

MacKenzie, James, Roger Dower, and Donald Chen. *The Going Rate: What It Really Costs to Drive*. Washington, DC: World Resources Institute, 1992.

Moreno, Rene and D.G. Bailey, *Alternative Transport Fuels from Natural Gas*. World Bank Technical Paper #98. Washington, DC: World Bank Publications, June 1989.

Morgan, Daniel. "Hydrogen as a Fuel." Washington, DC: Congressional Research Service, March 22, 1993.

Nadis, Steve and James MacKenzie. *Car Trouble*. Boston, Massachusetts: Beacon Press, 1993.

National Hydrogen Association. *The Hydrogen Technology Assessment*. (Prepared for the National Aeronautics and Space Administration). Washington, DC: 1992.

National Petroleum Council. *The Potential for Natural Gas in the United States*. Washington, DC: December 17, 1992.

National Renewable Energy Laboratory, US Department of Energy. *Proceedings: DOE/SERI Hydrogen Program Review, January 23-24, 1991*. Golden, Colorado: SERI, 1991.

National Renewable Energy Laboratory, US Department of Energy. *Proceedings: DOE/NREL Hydrogen Program Review, May 4-6, 1993.* (NREL/CP-470-5777) Golden, Colorado: NREL, August 1993. Annual program reviews were also published for 1991 and 1992.

Natural Gas Vehicle Coalition. *Strategic Plan: June 1, 1990.* Washington, DC: 1990.

Ogden, Joan and Robert Williams. *Solar Hydrogen: Moving Beyond Fossil Fuels.* Washington, DC: World Resources Institute, 1989.

Oppenheimer, Ernest. *Natural Gas: The Best Energy Source.* New York: Pen & Podium, Inc., 1989.

Renner, Michael. *Rethinking the Role of the Automobile.* Washington, DC: Worldwatch Institute, June 1988.

Riley, Robert. *Alternative Cars in the 21st Century: A New Personal Transportation Paradigm.* Warrendale, Pennsylvania: Society of Automotive Engineers, 1994.

Samsa, Michael. *Potential for Compressed Natural Gas Vehicles in Centrally-Fueled Automobile, Truck and Bus Fleet Applications.* Chicago, Illinois: Gas Research Institute, June 1991.

Santini, D.J., *et al., Greenhouse Gas Emissions from Selected Alternative Transportation Fuel Market Niches.* United States Department of Energy Center for Transportation Research, Argonne National Laboratory, Argonne, Illinois: 1989.

Schiffer, Michael. *Taking Charge: The Electric Automobile in America.* Washington, DC: Smithsonian Institution Press, 1994.

Skelton, Luther. *The Solar-Hydrogen Energy Economy.* New York: Van Nostrand Reinhold Co., 1984.

Sperling, Daniel (editor). *Alternative Transportation Fuels.* New York: Quorum Books, 1989.

Sperling, Daniel. *The Future Drive: Electric Vehicles and Sustainable Energy.* Washington, DC: Island Press, 1995.

Sperling, Daniel. *New Transportation Fuels: A Strategic Approach to Technological Change*. Berkeley, California: University of California Press, 1988.

Sperling, Daniel and Susan Shaheen (editors). *Transportation and Energy: Strategies for a Sustainable Transportation System*. Washington, DC: American Council for an Energy-Efficient Economy, 1995.

Steinhart, Carol and John. *Energy: Sources, Use, and Role in Human Affairs*. North Scituate, Massachusetts: Duxbury Press 1974.

Swan, David and Debbi Smith. *Hydrogen-Fueled Vehicles: Technology Assessment Report*. Prepared for and published by the California Energy Commission. Sacramento, California: June 29, 1991.

US Department of Energy. *Alternative Fuel Vehicles for the State Fleets: Results of the 5-Year Planning Process*. Washington, DC: May 1993.

US Department of Energy. *An Assessment of the Natural Gas Resource Base of the United States*. Washington, DC: May 1988.

US Department of Energy. *Assessment of Costs and Benefits of Flexible and Alternative Fuel Use in the US Transportation Sector*. Washington, DC: Four reports published 1988-1990.

US Department of Energy. *Hydrogen Program Plan: FY 1993 - FY 1997*. Golden, Colorado: National Renewable Energy Laboratory, June 1992. A companion *Hydrogen Program Implementation Plan: FY 1994 - FY 1998* was published in October 1993.

US Department of Energy, Office of Transportation Technologies. *Feasibility Study of Onboard Hydrogen Storage for Fuel Cell Vehicles: Interim Report*. Washington, DC: January 1993.

US Department of Energy, Office of Transportation Technologies. *National Program Plan: Fuel Cells in Transportation*. Washington, DC: February 1993.

US Energy Information Administration. *Alternatives to Traditional Transportation Fuels*. Washington, DC: June 1994.

US Energy Information Administration. *Annual Energy Review 1993*. Washington, DC: 1994.

US Energy Information Administration. *Emissions of Greenhouse Gases in the United States: 1987-1992*. Washington, DC: 1994.

US Energy Information Administration. *International Energy Annual 1992*. Washington, DC: 1994.

US Environmental Protection Agency. *Air Quality Benefits of Alternative Fuels*. Ann Arbor, Michigan: Prepared for the Alternative Fuels Working Group of the President's Task Force on Regulatory Relief, July 1987.

US Environmental Protection Agency. *Analysis of the Economic and Environmental Effects of Compressed Natural Gas as a Vehicle Fuel*. Ann Arbor, Michigan: April 1990.

US Environmental Protection Agency. *National Air Quality and Emissions Trends Report, 1991*. Report # 450-R-92-001, Research Triangle Park, North Carolina: October 1992.

US General Accounting Office. *Alternative Fuels: Experiences of Brazil, Canada and New Zealand in Using Alternative Motor Fuels*. GAO/RCED-92-119. Washington, DC: May 1992.

Wilson, B. *Evaluation of Aftermarket Fuel Delivery Systems for Natural Gas and LPG Vehicles*. NREL/TP-420-4892. Golden, Colorado: National Renewable Energy Laboratory, September 1992.

Zuckerman, Wolfgang. *End of the Road: The World Car Crisis and How We Can Solve It*. Post Mills, Vermont: Chelsea Green Publishing Co., 1991.

Index

A

B

C

D

E

F

G

I

J

O

P

P *(continued)*

R

S

About the Author

James S. Cannon, senior fellow at INFORM, is an internationally recognized researcher, author, and analyst on energy development, environmental protection, and related public policy issues. Long associated with INFORM, he is the author of *Drive for Clean Air* (1989), the influential study of natural gas and methanol as alternative vehicle fuels, and *Paving the Way to Natural Gas Vehicles* (1993), which identifies the obstacles to increased use of natural gas vehicles and ways to overcome those barriers.

Mr. Cannon has held employee and consultant positions with several not-for-profit organizations, private companies, and federal and state agencies. He had a seven-year association with the US Office of Technology Assessment, and for eight years was an Energy Policy Analyst for the New Mexico Energy, Minerals, and Natural Resources Department. Mr. Cannon is frequently called upon to testify on energy and environmental issues before government, academic, and public audiences.

Prior to his research on transportation issues, Mr. Cannon wrote two INFORM studies on acid rain, *Controlling Acid Rain: A New View of Responsibility* (1987) and *Acid Rain and Energy: A Challenge for New Jersey* (1984), as well as studies of coal conversion options in New York and New Jersey. He was co-author of INFORM's *A Clear View: Guide to Industrial Pollution Control* (1975) and of its 1976 report on alternative energy sources, *Energy Futures: Industry and the New Technologies.*

James Cannon received an A.B. in chemistry from Princeton University and an M.S. in biochemistry from the University of Pennsylvania.

Publications and Membership

Energy and Air Quality

Harnessing Hydrogen: The Key to Sustainable Transportation (James S. Cannon), 1995, 358 pp., $30.

Paving the Way to Natural Gas Vehicles (James S. Cannon), 1993, 192 pp., $25.

"Reformulated Gasoline: Cleaner Air on the Road to Nowhere" (James S. Cannon), 1994, 14 pp., $5.

Sustainable Products and Practices

Making Less Garbage on Campus: A Hands-On Guide (David Saphire), 1995, 72 pp., $20.

Less Garbage Overnight: A Waste Prevention Guide for the Lodging Industry (John P. Winter and Sharene L. Azimi), available early 1996, $15.

Delivering the Goods: Benefits of Reusable Shipping Containers (David Saphire), 1994, 30 pp., $20.

Case Reopened: Reassessing Refillable Bottles (David Saphire), 1994, 351 pp., $25.

Germany, Garbage, and the Green Dot: Challenging the Throwaway Society (Bette K. Fishbein), 1994, 276 pp., $25.

Making Less Garbage: A Planning Guide for Communities (Bette K. Fishbein and Caroline Gelb), 1993, 192 pp., $30.

Business Recycling Manual (copublished with Recourse Systems, Inc.), 1991, 202 pp., $42.50.

Burning Garbage in the US: Practice vs. State of the Art (Marjorie J. Clarke, Maarten de Kadt, Ph.D., and David Saphire), 1991, 288 pp., $47.

Reducing Office Paper Waste (Robert Graff and Bette Fishbein), 1991, 28 pp., $15.

Garbage Management in Japan: Leading the Way (Allen Hershkowitz, Ph.D., and Eugene Salerni, Ph.D.), 1987, 152 pp., $5.

Chemical Hazards Prevention

Toxics Watch 1995, 1995, 816 pp., $125.

Stirring Up Innovation: Environmental Improvements in Paints and Adhesives (John Young, Linda Ambrose, and Lois Lobo) 1994, 128 pp., $25.

A Clearer View of Toxics: New Jersey's Reporting Requirements as a Model for the United States (1994, 112 pp., $15)

Preventing Industrial Toxic Hazards: A Guide for Communities (Marian Wise and Lauren Kenworthy), 1993, 208 pp., $25.

Environmental Dividends: Cutting More Chemical Wastes (Mark H. Dorfman, Warren R. Muir, Ph.D., and Catherine G. Miller, Ph.D.), 1992, 288 pp., $75.

Cutting Chemical Wastes: What 29 Organic Chemical Plants Are Doing to Reduce Hazardous Wastes (David J. Sarokin, Warren R. Muir, Ph.D., Catherine G. Miller, Ph.D., and Sebastian R. Sperber), 1986, 548 pp., $47.50.

Tackling Toxics in Everyday Products: A Directory of Organizations (Nancy Lilienthal, Michèle Ascione, and Adam Flint), 1992, 192 pp., $19.95.

Toward a More Informed Public: Recommendations for Improving the Toxics Release Inventory (Jacqueline B. Courteau and Nancy Lilienthal), 1991, 26 pp., $10.

Sales Information

Payment

Payment, including shipping and handling charges, must be made in US funds drawn on a US bank and must accompany all orders. Please make checks payable to INFORM and mail to:

INFORM, Inc.
120 Wall Street
New York, NY 10005-4001

MasterCard and Visa are also accepted. Please call 212-361-2400 ext. 221 for more information.

Please include a street address; UPS cannot deliver to a box number.

Shipping Fees

United States: Add $3 for first book + $1 for each additional book. (4th class delivery; allow 4-6 weeks)
Canada: Add $5 for first book + $3 each additional book.
Foreign/surface: Add $8 for first book + $4 each additional book.
Foreign/airmail: Add $20 for first book + $10 each additional book.
Outside the US: Allow additional shipping time.

Priority shipping is higher; please call for charges.

Discount Policy

Booksellers: 20% on 1-4 copies of same title; 30% on 5 or more copies of same title

General bulk: 20% on 5 or more copies of same title

Students with ID; public interest and community groups with nonprofit federal ID:
Books under $10: no discount
Books $10-$25: cost $10
Books $25 and up: cost $15
Books $100 and up: cost $20

Government, upon request:
Books $45 and under: no discount
Books over $45: cost $45

Returns

Booksellers may return books, in saleable condition, for full credit or cash refund up to 6 months from date of invoice. Books must be returned prepaid and include a copy of the invoice or packing list showing invoice number, date, list price, and original discount.

Membership

Donors of US$25. or more will receive a one-year subscription to INFORM's quarterly newsletter, *INFORM Reports,* and advance notification of all INFORM publications.

In addition, donors of gifts of US$1,000. or more may be offered complimentary copies of INFORM publications. Acceptance of these publications may reduce the amount of the donor's tax deduction.

For residents of Canada, our basic membership begins at US$35; all other countries US$50. All payments must be made in US funds drawn on a US bank.

INFORM Board of Directors